# A GENERAL INTRODUCTION TO FRACTURE MECHANICS

# A GENERAL INTRODUCTION TO FRACTURE MECHANICS

*A Journal of Strain Analysis Monograph*

MECHANICAL ENGINEERING PUBLICATIONS LIMITED
LONDON

*First Published 1978*

*This publication is copyright under the Berne Convention and the International Copyright Convention. Apart from any fair dealing for the purpose of private study, research, criticism, or review, as permitted under the Copyright Act 1956, no part may be reproduced, stored in a retrieval system or transmitted in any form or by any means, electronic, electrical, chemical, mechanical, photocopying, recording or otherwise, without the prior permission of the copyright owners.*

© *The Institution of Mechanical Engineers 1978*

*ISBN 0 85298 383 2*

Printed by The Burlington Press, Foxton, Herts.

# Contents

|   |   | PAGE |
|---|---|---|
| General notation | | vii |
| Foreword | | viii |
| 1. | Origins of the energy balance approach to fracture<br>*D. J. Hayes* | 1 |
| 2. | Origins of the stress intensity factor approach to fracture<br>*D. J. Hayes* | 9 |
| 3. | The fracture toughness of metals<br>*J. F. Knott* | 17 |
| 4. | Yielding fracture mechanics<br>*C. E. Turner* | 32 |
| 5. | Evaluation of stress intensity factors<br>*D. J. Cartwright and D. P. Rooke* | 54 |
| 6. | Experimental methods for fracture toughness measurement<br>*A. H. Priest* | 74 |
| 7. | Application of fracture mechanics to the brittle fracture of structural steels<br>*R. R. Barr and P. Terry* | 93 |
| 8. | Analysis and application of fatigue crack growth data<br>*L. P. Pook* | 114 |
| 9. | Environmental effects in crack growth<br>*R. N. Parkins* | 136 |
| 10. | Fracture mechanics: a summary of its aims and methods | 153 |
| General Reference Section | | 156 |
| Index | | 177 |

Published by Mechanical Engineering Publications Ltd., London and New York for the Institution of Mechanical Engineers (Tel: 01-839 1211) on behalf of The Joint British Committee for Stress Analysis.

*Editorial Office:* P.O. Box 24, Northgate Avenue, Bury St Edmunds, Suffolk IP32 6BW. Tel: 0284 63277.

# General notation

(Adapted from Draft British Standard DD3 and DD19. Other symbols are defined as they occur.)

| | |
|---|---|
| $a$ | Crack length, half length of internal crack |
| $B$ | Test piece thickness |
| $E$ | Young's modulus |
| $G$ | Strain energy release rate per unit thickness with crack advance |
| $J$ | $J$ contour integral |
| $K$ | Stress intensity factor, subscript I, II, III denote mode* |
| $K_c$ | Critical value of $K$ for crack growth |
| $K_f$ | Maximum value of $K_I$ during final stage of fatigue precracking |
| $K_Q$ | Provisional value of $K_{Ic}$ |
| $K_{Ic}$ | Value of $K_c$ under plane strain conditions |
| $\Delta K$ | Range of $K$ during fatigue cycle |
| $P$ | Potential energy |
| $Q$ | Applied force |
| $Q_5$ | Value of $Q$ at intersection of secant line with force displacement curve |
| $Q_Q$ | As for $Q_5$ or any higher preceding force |
| $Q_c$ | Critical value of $Q$ for crack growth |
| $q$ | Displacement |
| $S$ | Loading span |
| $V_g$ | Clip gauge displacement |
| $W$ | Test piece width |
| $Y$ | Stress intensity factor coefficient* |
| $z$ | Distance of clip gauge from test piece surface |
| $\gamma$ | Specific surface energy, or surface tension |
| $\delta$ | Crack opening displacement |
| $\delta_c$ | Critical value of $\delta$ for crack growth |
| $\mu$ | Shear modulus |
| $\nu$ | Poisson's ratio |
| $\sigma$ | Stress |
| $\sigma_Y$ | Yield stress (taken as 0·2 per cent proof stress) |

* The numerical value of $K$ is represented by some authors as $Y\sigma\sqrt{a}$ and by others as $Y\sigma\sqrt{(\pi a)}$ the value of $Y$ in the former usage being equal to that of $Y\sqrt{\pi}$ in the latter. In this group of papers the formula $K = Y\sigma\sqrt{a}$ is used in Chapters 2 and 8 and the formula $K = Y\sigma\sqrt{(\pi a)}$ in Chapters 3, 5, and 7; elsewhere the value of $Y$ is not brought into the discussion.

# Foreword

This Special Issue of the Journal of Strain Analysis originated in proposals made by the Editorial Board. A need was recognized for an up-to-date review of the main topics of modern fracture mechanics, written in textbook style, but in a format which would be cheap enough for general use by students. A group of specialists were invited to write papers to a predetermined plan with material presented in logical rather than historical order. The papers are divided into two groups, the first giving the theoretical background to various fracture mechanics concepts, and the second concentrating on its practical aspects.

A convenient and reasonably rigorous definition of fracture mechanics is 'the applied mechanics of crack growth'. Thus fracture mechanics does not tell us anything about fracture processes, but it does provide the necessary descriptive and analytic framework for their study. Present day fracture mechanics deals largely with macroscopic aspects of crack growth. Fracture surfaces are assumed to be smooth although microscopically they are actually very irregular. In the course of analysis several other simplifying assumptions are also made; for instance, the development of the important concept of stress intensity factor has been based on the assumption that the material is a linear elastic continuum. Various modifications to basic theory can be made to account for the actual behaviour of real materials, and much recent work is concerned with the rigorous analysis of situations involving gross plasticity. In early fracture mechanics semi-empirical derivations often had to be used because rigorous solutions were not available. This sometimes led to a lack of confidence in the utility of fracture mechanics concepts in practical engineering, but the situation is now much improved.

Fracture mechanics has helped to quantify the rather elusive concept of 'toughness'. It can now be usefully defined as 'resistance to crack growth'. Notice that the type of loading (static, fatigue, etc.) and environment are not specified. Historically, early work was on brittle fracture, spurred on by the Liberty ship failures at the end of the Second World War, the Comet disasters and problems with space rockets. However the development of the important concept of stress intensity factor $K$, as a single parameter description of the elastic stress field in the vicinity of a crack tip, has encouraged the application of fracture mechanics to virtually any type of crack growth. Notice that fracture mechanics deals only with the cracked situation; it cannot help with situations involving uncracked materials. However the majority of structures contain cracks, which are either introduced during the manufacture, or are initiated early in the life of the structures, and these are frequently the source of service failures.

The logical basis of fracture mechanics theory for static loadings is the premise that for crack growth to occur two conditions are necessary and sufficient. Firstly, sufficient stress must be available at the crack tip to operate some mechanism of crack growth and, secondly, sufficient energy must flow to the crack tip to supply the work done in the creation of new surfaces. Initially it was believed that only the first was required; however Inglis' solution for an elliptical hole indicated that the elastic stress tends to infinity as a crack tip is approached, leading to the paradoxical conclusion that a cracked body cannot support any load. This paradox was resolved by Griffith, who used an energy balance approach based on surface energy to explain the fracture behaviour of glass. The energy balance approach was extended by Orowan and Irwin to include the energy associated with plastic deformation adjacent to the new crack surfaces, and more recently by Rice and others to situations involving gross plasticity.

Meanwhile Irwin had introduced the concept of stress intensity factor and had shown that crack growth occurring at a critical value was equivalent to the satisfaction of both conditions. The use of the concept has the great advantage that material properties in the presence of a crack can be measured in terms of stress intensity factors in just the same way that the material properties of a plain specimen are measured in terms of stress. However the use of stress intensity factor (or any other fracture mechanics parameter) for the measurement of properties can only be defended on the empirical grounds that adequately consistent results are obtained when stated rules are followed, and of convenience, utility and sufficient accuracy for a particular application. It is on these grounds that modern fracture mechanics methods are now used in a wide range of problems, particularly those involving fatigue crack growth and brittle fracture.

## PLANNING PANEL

H. L. Cox, Chairman
Professor H. Fessler
Dr K. J. Miller
Dr L. P. Pook
Professor C. E. Turner

# 1
# Origins of the Energy Balance Approach to Fracture

D. J. HAYES

*Shell Research BV, Koninklijke Shell-Laboratorium, Amsterdam*

The development of the energy balance approach to fracture is traced. Limitations of the approach are discussed and the justification for its application to practical problems is indicated.

The energy balance approach to the study of the fracture phenomenon in cracked bodies was originally proposed by Griffith (1/1)*. His basic premise was that unstable propagation of a crack takes place if an increment of crack growth results in more stored energy being released than is absorbed by the creation of the new crack surface. This statement is appealing since it conforms to elementary ideas as to the requirements for an unstable process. However, when considered in detail the evaluation of these energy terms presents great difficulties in most practical situations. For example, if crack extension takes place after considerable plastic deformation it is not a simple task to evaluate the energy that is released by a crack extension. These difficulties tended to direct attention towards crack tip characterizing parameters as measures of the susceptibility of materials to fracture. Such approaches are the subject of companion papers and hence are not pursued here.

The difficulties of interpretation mentioned above do not occur if attention is restricted to (hypothetical) materials which behave in a purely elastic manner prior to crack propagation and where all absorbed energy is associated with the classical interpretation of surface energy. Griffith (1/1)

---
*The MS of this paper was received at the Institution of Mechanical Engineers on 30 April 1975 and accepted for publication on 22 August 1975.*
\* References are given in the general reference section on page 156.

confined himself to consideration of this problem. As a specific case he dealt with a crack, of length $2a$, contained within a plane elastic body of indefinite extent. Remote from the crack uniform direct stresses acted perpendicular and parallel to the crack plane. These stresses are denoted as $\sigma_{yy}^{(\infty)}$ and $\sigma_{xx}^{(\infty)}$ respectively. This situation is illustrated schematically in Fig. 1. The total strain energy of such a body is perhaps most readily determined by making use of a result due to Bueckner (**1/2**). This result demonstrates that the strain energy associated with the introduction of the crack in the body shown in Fig. 1 is identical with the strain energy of the body

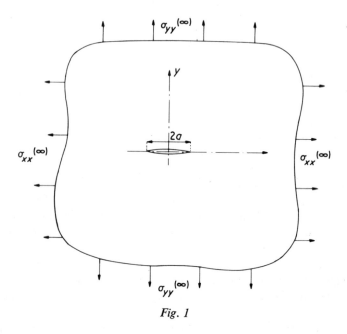

Fig. 1

shown in Fig. 2. In this case locations remote from the crack are stress-free and a uniform stress $(\sigma_o)$, equal to $\sigma_{yy}^{(\infty)}$, is applied perpendicular to the crack faces. This implies that, under the restrictions considered, the stress $\sigma_{xx}^{(\infty)}$ has no influence on crack stability. For this situation the deformed surface of the upper crack face is given by (**1/3**):

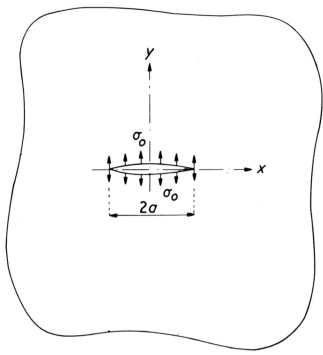

Fig. 2

$$u = -\frac{\sigma_0 x}{4\mu}(\kappa - 1)$$
$$v = \frac{\sigma_0}{4\mu}(\kappa - 1)\sqrt{(a^2 - x^2)}\quad\bigg\}\quad |x| \leqslant a$$

where:

$\mu$ is the shear modulus
$\kappa$ is a function of Poisson's ratio ($v$)
   viz. $\kappa = 3 - 4v$ for plane strain
   $= \dfrac{3-v}{1+v}$ for plane stress

and $u$, $v$ are components of displacement in the coordinate directions $x$, $y$ which are respectively parallel and perpendicular to the crack plane, with the origin corresponding to the mid-point of the crack.

The strain energy of the body shown in Fig. 2 is equal to the work done in deforming the crack surfaces. This is given by:

$$U = 4 \times \frac{1}{2} \int_0^a \frac{\sigma_o^2 (\kappa + 1)}{4\mu} \sqrt{(a^2 - x^2)} \, t \, dx$$

or:

$$U = \frac{(\kappa + 1)}{8\mu} \pi a^2 \, t \sigma_o^2$$

where $t$ is the (constant) thickness of the body. This result is identical with the corrected result given by Griffith in his second paper (1/4).

Returning to the original problem, the total stored energy of the body shown in Fig. 1 is given by:

$$U_{tot} = \frac{(\kappa + 1)\pi a^2 \, t \sigma_{yy}^{(\infty)}}{8\mu} + \overline{U}$$

where $\overline{U}$ is the component of energy which is not dependent on the presence of the crack. The total surface energy of the crack, for the case considered, is given by:

$$U_s = 4at\gamma$$

where $\gamma$ is the specific surface energy or surface tension.

If an increment in crack length, $2\delta a$, takes place the change in strain energy is given by:

$$\delta U_{tot} = \frac{(\kappa + 1)\pi \, a \, t \, \sigma_{yy}^{(\infty)2}}{4\mu} \cdot \delta a$$

whilst the increase in surface energy is:

$$\delta U_s = 4 \, t \, \gamma \cdot \delta a$$

Thus, according to the basic premise of Griffith (1/1), crack growth will be unstable if:

$$\frac{(\kappa + 1)\pi \, a \, t \, \sigma_{yy}^{(\infty)2}}{4\mu} \cdot \delta a > 4t\gamma \cdot \delta a$$

or:

$$\sigma^* \geqslant 4\sqrt{\left(\frac{\mu\gamma}{\pi a(\kappa+1)}\right)}$$

where $\sigma^*$ is the value of $\sigma_{yy}^{(\infty)}$ required to cause instability. Therefore, under conditions of plane strain unstable fracture will take place if:

$$\sigma^*\sqrt{a} \geqslant \sqrt{\left(\frac{2E\gamma}{\pi(1-v^2)}\right)}$$

whilst for plane stress this instability occurs when:

$$\sigma^*\sqrt{a} \geqslant \sqrt{\left(\frac{2E\gamma}{\pi}\right)}$$

where $E$ is Young's modulus.

In order to test his premise Griffith carried out a series of experiments on hard glass. He argued that there were grounds to expect that the surface tension of glass would be a linear function of temperature. Thus, from a series of ancillary tests on glass at elevated temperature he was able to estimate the value of $\gamma$ at ambient temperature. Under normal conditions glass exhibits minute plastic distortion prior to fracture. The strain energy release term for the cracked body as computed above should therefore be valid for interpreting the fracture of glass specimens provided the cracks may be regarded as lying in an infinite plane body. Griffiths chose to test to failure a number of cracked glass tubes and spheres. The results of these experiments were very encouraging; at the instant of fracture, a constant value of $\sigma^*\sqrt{a}$ was obtained for a range of crack lengths. Further stresses applied parallel to the crack plane had no influence on the value of $\sigma^*\sqrt{a}$. The correlation of results according to the predicted relation:

$$\sigma^* = \sqrt{\left(\frac{2E\gamma}{\pi a}\right)}$$

was not entirely satisfactory however†.

---

† Griffith (1/1) found excellent agreement between $\sigma^*$ and $\sqrt{\left(\frac{2E\gamma}{v\pi a}\right)}$.
However, the inclusion of $v$ was due to a calculation error. When this error is removed the measured value of $\gamma$ was low by a factor of 3 in terms of the fracture theory. As stated in his paper: '. . .some reconsideration of the experimental verification of the theory is necessary'.

From the practical viewpoint the demonstration that a functional relationship exists between failure stress and crack length was a significant step forward. In fact this ability. albeit in limited circumstances, to predict the fracture behaviour of one cracked body on the basis of the observed behaviour of another may be regarded as the object of fracture mechanics.

Strictly the work of Griffith (1/1, 1/4) may only be applied to materials where non-linear effects, prior to fracture, are absent on the continuum scale. This restriction virtually rules out consideration of engineering situations. However, almost thirty years after Griffith's contribution, Irwin (1/5) and Orowan (1/6) suggested a modification to the original formulation so that limited plastic deformation prior to failure could be accommodated by the theory. Their approach was to replace the surface energy term $2\gamma$ by a term $\gamma_p$ which represented the energy of plastic distortion absorbed by the fracture process. Orowan (1/7) noted that this plastic energy term was approximately three orders of magnitude greater than the surface energy term and, hence, if a Griffith type energy balance approach is appropriate, this latter term may be neglected. Both Irwin and Orowan argued that, provided plastic distortion takes place in a zone which is small in comparison with crack length and component thickness, the energy released by crack extension could still be calculated from elastic analysis with a sufficient degree of accuracy. Thus, in essence, the modified theory simply involves a redefinition of the energy absorption term.

Orowan suggested that strict conditions should be placed on situations where the modified Griffith theory could be applied. He pointed out (1/7) that, strictly speaking, the approach would only be valid if plastic deformation was confined to thin layers adjacent to the crack walls. In a series of experiments Felbeck and Orowan (1/8) demonstrated that fracture stress was inversely proportional to the square root of crack length, as predicted by the modified theory. However, the value of $\gamma_p$ calculated from these results was 2 to 5 times that to be expected from X-ray analysis of the fracture surfaces. They concluded that these results would be better explained using a crack tip characterizing parameter approach‡.

Irwin's explanation of the modified theory was more pragmatic than that of Orowan. He argued (1/9) that a precise interpretation of the plastic surface energy term was unnecessarily restrictive. Provided the plastic zone was small, a theory for correlating fracture behaviour could be substanti-

‡ This approach was also followed by Griffith in his second paper (1/4) possibly for similar reasons.

ated. In his view the modified theory consisted in evaluating the rate of strain energy release at the point of fracture. If the fracture process was essentially similar for different loadings and geometries, the fracture event would occur when the strain energy release rate reached a critical value. This critical value could be regarded as a material property to be determined by a fracture test.

A means for determining the value of energy release rate for different loading conditions and geometries was proposed by Irwin and Kies (1/9). They noted that the strain energy in an elastic body could be represented by the relationship

$$U = \frac{Q^2 c}{2}$$

In this relationship $Q$ is a characterizing force and $c$ the compliance of the body, i.e. the displacement at the point of application of $Q$ due to unit $Q$. From this expression it immediately follows that the strain energy release rate, with respect to crack extension, is given by

$$\frac{\partial U}{\partial a} = \frac{1}{2} Q^2 \frac{\partial c}{\partial a}$$

Irwin and Kies (1/9) suggested that by measuring the compliance of a test specimen, or a component model, with various crack lengths the value of $\partial c/\partial a$ as a function of crack length could be obtained. A fracture test could then be interpreted by evaluating $\partial U/\partial a$ at fracture using the fracture load and the value of $\partial c/\partial a$ for the appropriate crack length. On the basis of this result a considerable amount of work was carried out to determine the compliance of various specimen shapes; for example, Day (1/11) and Srawley *et al* (1/12). Although later developments have to a certain extent superseded the compliance method, in some circumstances it still retains its utility.

Irwin (1/13) proposed an alternative interpretation of critical strain energy release rate when he suggested that this quantity could be regarded as a force. In these terms fracture was described as a rate-controlled process driven by this force, which was defined as the irreversible energy loss per unit area of newly created surface. This force, denoted as $G$ (after Griffith) would have a critical value, $G_c$, when a crack starts to propagate. As stated by Irwin (1/14), 'The $G_c$ concept appears to have about the same justification in relation to fracture as pertains to the yield strength concept in the case of plastic deformation'.

In situations where fracture is preceded by limited plastic deformation there is a strict equivalence between the strain energy release rate concept, $G$, and the stress intensity factor approach§. Thus the techniques available for determining stress intensity factors are equally valid for determining $G$. However, in situations where extensive plastic deformation takes place prior to failure the relationship between an energy balance approach and a crack tip environment approach becomes more tenuous. There is therefore a need to exercise great care when interpreting fracture behaviour in these situations.

To summarize the concept of energy balance, as originally proposed by Griffith and extended by Irwin and Orowan, the following points are made. It is postulated that unstable crack propagation takes place when the strain energy released in an incremental crack extension exceeds the energy that is absorbed by creating new crack surface. The determination of the strain energy released is a relatively straight-forward matter only when non-linear deformations take place in a region of limited extent in the immediate vicinity of a crack tip. Precise determination of absorbed energy from independent measurement does not provide a satisfactory means of predicting the fracture event. However, the practical approach proposed by Irwin does enable the onset of unstable fracture to be predicted in real structures provided suitable fracture experiments on cracked specimens have been carried out. This ability to predict failure is sufficient justification for utilizing the concept as an engineering approach.

§ This approach is discussed in chapter 2, Origins of the stress intensity factor approach to fracture.

# 2
# Origins of the Stress Intensity Factor Approach to Fracture

D. J. HAYES

*Shell Research BV, Koninklijke/Shell Laboratorium, Amsterdam*

The basic philosophy of the stress intensity approach to fracture is scrutinised. The limitations of the approach are discussed and its application to the solution of practical problems is indicated.

In chapter 1 the development of fracture mechanics on the basis of an energy balance approach was discussed. An alternative interpretation of fracture phenomena which focuses attention on the mechanical environment near the tip of a crack will now be considered. This interpretation was originally developed by Irwin (2/1, 2/2) and is generally known as the stress intensity factor approach.

By considering the analyses due to Westergaard (2/3), Irwin (2/1, 2/2) noted that, for the cases given, the stresses in the vicinity of a crack tip could be expressed in the following form:

$$\begin{bmatrix} \sigma_{xx} \\ \sigma_{yy} \\ \tau_{xy} \end{bmatrix} = \frac{K \cos\frac{\theta}{2}}{\sqrt{(2\pi r)}} \begin{bmatrix} 1 - \sin\frac{\theta}{2}\sin\frac{3\theta}{2} \\ 1 + \sin\frac{\theta}{2}\sin\frac{3\theta}{2} \\ \sin\frac{\theta}{2}\cos\frac{3\theta}{2} \end{bmatrix}$$

$+$ terms of order $r^0$

*The MS of this paper was received at the Institution of Mechanical Engineers on 30 April 1975 and accepted for publication on 22 August 1975.*

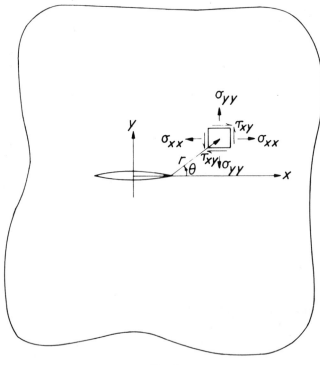

*Fig. 3*

where $r$, $\theta$ are the cylindrical polar coordinates of a point with respect to the crack tip (see Fig. 3).

Thus a characteristic spatial distribution of stresses was found, each specific case being characterized by the stress intensity factor, $K$. Further, on the basis of an elastic analysis, stresses become infinitely large as the crack tip is approached. This necessity of dealing with singular terms may lead to conceptual difficulties. However, such difficulties can be rationalized provided it is realized that below a certain scale continuum analysis is inappropriate and that such analyses provide descriptions of boundary conditions on these regions.

By using virtual work arguments Irwin (**2/1, 2/2**) was able to demonstrate that the strain energy release rate or crack extension force, $G$, (see chapter 1) could be identified with $K$ according to the following relationships:

$$G = \frac{1+\kappa}{8\mu} K^2$$

where $\mu$ is the shear modulus and $\kappa$ is a function of Poisson's ratio ($\nu$)

viz $\kappa = 3 - 4\nu$ for conditions of plane strain

$\quad\ = \dfrac{3-\nu}{1+\nu}$ for conditions of plane stress

This relationship shows that if fracture (preceded by limited plastic flow) may be characterized by the attainment of a critical crack extension force then this is equivalent to characterizing the fracture event by the attainment of a critical stress environment. Equally, since elasticity is the basis for these analyses, the near-tip environment may be thought of in terms of strain, displacement, strain energy density*, etc. In all these cases the factor $K$ enters as a multiplicative constant and a unique angular distribution is associated with each field parameter. Thus $K$ should not be regarded as being particularly associated with stress. Rather, it should be interpreted as a parameter characterizing the mechanical environment as a whole. Further, for a predominantly elastic situation the concept of a critical local crack tip environment controlling fracture is entirely equivalent to the concept of a critical force driving the crack to extend or to a critical strain energy release rate just sufficient to supply the required energy for the fracture process. In short, fracture in these circumstances can be characterized by the attainment of a critical value of $K$†.

The results available in Westergaard's paper (2/3) related to situations where a single crack, or an infinite array of regularly spaced collinear cracks, were subject to loading normal to the crack plane. Subsequent papers (2/5–2/16) demonstrated, both for particular problems and for classes of problems (two-dimensional in-plane and anti-plane loading, three-dimensional loading, plate bending and shell problems), that the characteristic distribution of elastic field quantities in the vicinity of a crack tip always resulted. For the case of a rectilinear crack lying in the $x$–$z$ plane (see Fig. 3) the results may be summarized as:

---

* Strain energy density in particular is receiving considerable attention as a characterizing quantity for general situations (2/4).
† Phenomena such as fatigue crack growth and stress corrosion cracking can also be rationalized in terms of the characterizing parameter $K$. See chapters 8 and 9.

$$\begin{bmatrix} \sigma_{xx} \\ \sigma_{yy} \\ \tau_{xy} \end{bmatrix} = \frac{K_{\mathrm{I}} \cos\frac{\theta}{2}}{\sqrt{(2\pi r)}} \begin{bmatrix} 1 - \sin\frac{\theta}{2}\sin\frac{3\theta}{2} \\ 1 + \sin\frac{\theta}{2}\sin\frac{3\theta}{2} \\ \sin\frac{\theta}{2}\cos\frac{3\theta}{2} \end{bmatrix}$$

$$+ \frac{K_{\mathrm{II}} \sin\frac{\theta}{2}}{\sqrt{(2\pi r)}} \begin{bmatrix} 2 + \cos\frac{\theta}{2}\cos\frac{3\theta}{2} \\ \cos\frac{\theta}{2}\cos\frac{3\theta}{2} \\ \cos\frac{\theta}{2} - \sin\frac{3\theta}{2} \end{bmatrix}$$

$$+ \text{ terms of order } r^0$$

$$\begin{bmatrix} \tau_{xz} \\ \tau_{yz} \end{bmatrix} = \frac{K_{\mathrm{III}}}{\sqrt{(2\pi r)}} \begin{bmatrix} \sin\frac{\theta}{2} \\ \cos\frac{\theta}{2} \end{bmatrix} + \text{ terms of order } r^0,$$

the value of $\sigma_{zz}$ depending on whether the stress state is plane strain ($\epsilon_{zz} = 0$) or generalized plane stress ($\sigma_{zz} = 0$).

The subscripts that are given here characterize the three basic stress environments that are experienced on the crack plane. The subscript I refers to the case where the in-plane ($x$–$y$ plane) loading is symmetric with respect to the crack plane; subscript II refers to the case where the in-plane loading is skew-symmetric with respect to the crack plane, and subscript III refers to the case where the loading is anti-plane shear (shear loading in $x$–$z$ and $y$–$z$ planes). Since the above mentioned solution is developed from a linearly elastic analysis, the component parts of an arbitrary loading system can be decomposed to give the respective $K$'s. These $K$ parameters are termed 'stress intensity factors' and are functions only of geometry and loading conditions, provided the loading is in equilibrium.

Formally the relationship between $K$ and $G$ may be generalized to cover the three basic loading conditions, viz:

$$G_{\text{I}} = \frac{\kappa + 1}{8\mu} K_{\text{I}}^2$$

$$G_{\text{II}} = \frac{\kappa + 1}{8\mu} K_{\text{II}}^2$$

$$G_{\text{III}} = \frac{K_{\text{III}}^2}{2\mu}$$

However, under skew-symmetric and anti-plane loading conditions cracks tend to extend in a non-planar fashion. Hence, a criterion of fracture based on the attainment of critical values of $G_{\text{II}}$ or $G_{\text{III}}$ becomes difficult to justify. Indeed, the topic of non-planar crack extension is particularly difficult to treat and is still subject to controversy. Most practcal cases are concerned with loading that is symmetric with respect to the crack plane. In these circumstances only the variables with subscript I apply. If the reader restricts his attention to these cases he can avoid conceptual difficulties without his ability to deal with practical problems being impaired.

The philosophy of linear elastic fracture mechanics may be regarded as having been established when the equivalence of $G_{\text{I}}$ and $K_{\text{I}}$ had been demonstrated, [Irwin (**2/1, 2/2**)]. Elastic analysis had shown that the stress environment around a crack tip was entirely similar to all situations to within a linear scaling factor. By means of tests on suitably shaped and loaded specimens it was possible to determine the material property $K_{\text{Ic}}$ (or $G_{\text{Ic}}$) by defining it as the value of $K_{\text{I}}$ (or $G_{\text{I}}$) operative at the point of fracture. It was then possible to establish what flaws were tolerable in an engineering structure under given conditions or to compare materials as to their utility in situations where fracture is possible. However, it became apparent that, except in cases where fracture occurred at very low stress levels (in comparison with material yield stress), some account of plastic behaviour was necessary. The value of $K_{\text{I}}$ at the point of fracture was found to be strongly dependent on plate thickness, [Irwin (**2/17**)] and only after a certain thickness had been exceeded could the critical value be regarded as a material property, $K_{\text{Ic}}$ dependent only on the testing environment. The variation in the apparent value of $K_{\text{Ic}}$ has been attributed to the through-the-thickness change in constraint along the crack front. The plastic regions that are near a free surface are practically in a condition of plane stress whilst those remote from such a surface approach conditions of plane strain. When thickness is sufficient the fracture behaviour will be dominated by the region of constrained plastic deformation, a characteristic flat fracture will occur and conditions are described as 'plane strain'.

The recognition that plastic zones form at a crack tip prior to fracture has an important consequence for the interpretation of the stress intensity factor as a characterizing parameter. Liu **(2/18)** pointed out that it is not the elastic stresses and strains outside the plastic zone which cause fracture. Rather, the fracture behaviour is controlled by the mechanical environment within the plastic zone in the immediate vicinity of the crack tip. Within a certain radius the stress intensity factor characterizes the stress field and provided the plastic zone is sufficiently small compared to this radius the elastic field will be unaffected by plastic relaxation. Thus, for two different situations, say two different geometries, where the same stress intensity factor is applied conditions of stress, strain, etc at geometrically similar points will be identical even within the plastic zones provided these are much smaller than the radius at which conditions may be considered as specified by $K_I$. Under these restrictions the fracture event will be characterized by the attainment of $K_{Ic}$, the critical value of $K_I$.

As demonstrated by Liu **(2/18)**, and recently re-emphasized by Larsson and Carlsson **(2/19)**, the requirements for the conditions of similitude outlined above are very strict indeed. Thus, in practice, some relaxation of these requirements is tolerated in order that realistic assessments may be made.

So far the basic philosophy of the stress intensity approach has been discussed without mentioning precisely how specific cases are dealt with. Obviously, to interpret test results or to make design calculations, it is necessary to have explicit expressions for $K_I$ for specific geometries and loading conditions. The determinination of stress intensity factors is a specialist task necessitating the use of a number of analytical and numerical techniques. It would be inappropriate to discuss these methods here‡. The important point to note is that it is always possible to determine $K_I$ to a sufficient accuracy for any given geometry or set of loading conditions. In many instances solutions, suitable for preliminary calculations at least, are available in compendia **(2/12, 2/20)**.

In general stress intensity factors may be written in the form:

$$K_I = \sigma\sqrt{a}\, Y\left(\frac{a}{W}\right)$$

‡ For detailed discussion of these methods see chapter 5.

*a*    Isolated crack in an infinite plate

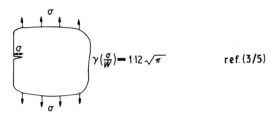

*b*    Surface crack in a semi-infinite plate

*c*    Surface crack in three-point bend specimen    ref (3/22)

*Fig. 4*

where $\sigma$ is a characterizing stress, $a$ is a characterizing crack length, $W$ is a characterizing dimension, and $Y(a/W)$ is a calibration function which defines $K_I$ for the specific body under consideration.

Figure 4 shows three configurations with the corresponding values of $Y(a/W)$. Thus, for example, a fracture test on a three-point bend specimen would be interpreted by substituting the appropriate value of $a/W$ in the polynomial for $Y(a/W)$ (Fig. 4c), noting the load at fracture and thus determining the value of $K_I$ at fracture. Alternatively, if it is necessary to estimate the maximum permissible depth of a surface defect in a thick component subject to uniform stress, use is made of the solution given in Fig. 4b. Here the stress intensity factor is given by:

$$K_I = 1 \cdot 12\, \sigma \sqrt{(\pi a)}$$

Consequently a knowledge of the critical value of $K_I$, as determined by an appropriate fracture test, and the applied stress level permits the critical value of crack depth to be determined.

To summarize, the underlying philosophy of the stress intensity factor approach to fracture is that the mechanical environment in the immediate vicinity of a crack tip has a unique distribution, independent of loading conditions or geometric configuration, provided non-linear behaviour is of limited extent. Geometry and loading conditions influence this environment through the parameter $K_I$, which may be determined by suitable analysis. The fracture event is interpreted as being characterized by the attainment of a critical value of the stress intensity factor, $K_{Ic}$. A knowledge of $K_{Ic}$, obtained from a suitable test, thus provides a means for predicting the fracture behaviour of real structures. The sensitivity of structures to other phenomena such as fatigue crack growth and stress corrosion cracking can likewise be predicted on the basis of suitably performed tests interpreted in terms of the stress intensity factor approach.

# 3
## The Fracture Toughness of Metals
### J. F. KNOTT
*Department of Metallurgy and Materials Science, University of Cambridge*

Problems in applying lefm to the fracture of metals are discussed. Fracture toughness values for cleavage and fibrous modes are related to the micromechanics of fracture.

### 3.1 INTRODUCTION

The importance of fracture mechanics is that it enables quantitative relationships to be obtained between the applied *stress* necessary to cause failure in a structure or testpiece and the *size of any defect* or precrack that may be present. The quantity which must be measured to link these two parameters is a material's *fracture toughness*. In an infinite body, containing a central, through-thickness crack of length $2a$, the Griffith equation (3/1) states that *elastic* fracture will occur at a critical value of the potential energy release rate per unit thickness, $G \equiv (1/B)(dP/da)$, given by:

$$G_{\text{crit}} = 2\gamma \tag{1}$$

where $2\gamma$ is the elastic 'work to fracture', $\gamma$ being the surface energy per unit area. Alternative forms of equation (1) make use of the relationship between stress intensity factor, $K$, and energy release rate, $G$ derived from the work done during virtual crack extension (see, for example (3/2)):

$$G = \alpha \frac{K^2}{E} \tag{2}$$

*The MS of this paper was received at the Institution of Mechanical Engineers on 25 May 1975 and accepted for publication on 22 August 1975.*

for mode I tensile loading; where $\alpha = 1$ in plane stress and $(1 - v^2)$ in plane strain; $v$ is Poissons ratio and $E$ is Young's modulus. Thus, an equivalent form of equation (1) is

$$K_{\text{crit}} = \sigma_F \sqrt{(\pi a)} \tag{3}$$

where $\sigma_F$ is the fracture stress of the infinite body containing a crack of length $2a$, or, by further substitution in (1).

$$\sigma_F = \sqrt{\left(\frac{EG_{\text{crit}}}{\pi \alpha a}\right)} \tag{4}$$

where $G_{\text{crit}}$ is equal to $2\gamma$ for an elastic fracture.

Most metals do not fail in a completely elastic manner, but only after some localized yielding has occurred at the crack tip. Then, provided that the non-elastic region is small compared with specimen and crack dimensions (*quasi-elastic fracture*), it is possible to calculate the potential energy release rate using elastic theory. *Experimentally*, it has been found that, for rather brittle metals, fracture is still characterized by a critical value of $G$, but that this is partially non-recoverable, stored for example, as strain energy in the dislocations present in the plastic zone, and only partially available to create new surface:

$$G_{\text{crit}} = 2\gamma + \gamma_p \tag{5}$$

where $\gamma_p$ is a plastic work term, usually found to be very much larger than $2\gamma$. To a first approximation, the failure stress is still given by equation (4) and the associated critical value of stress intensity, from equation (3), is then known as a material's *fracture toughness*. Mainly, it is a measure of the amount of plastic deformation preceding catastrophic crack propagation and so an understanding of a metal's toughness in a practical sense rests on explaining why a particular plastic zone size had to be attained before the crack could propagate.

## 3.2 FINITE BOUNDARIES

In the singular field immediately ahead of a crack tip, the stresses are conveniently expressed in the form:

$$\sigma = \frac{K}{\sqrt{(2\pi r)}} f(\theta) + \ldots \text{series} \tag{6}$$

where $r$ and $\theta$ are polar co-ordinates with origin at the crack tip. Equation (6) is the first term in a series expansion of a more general expression for the stress distribution and its form holds only for $r \ll a$. The stress intensity factor, $K$, is given by $K = \sigma_{app}\sqrt{(\pi a)}$* for the infinite body (cf the fracture criterion in equation (3)), but the presence of free surfaces at finite boundaries modifies the general distribution of stress and so alters the value of $K$ obtained by a given pair of values, $\sigma$ and $a$. For standard testpieces (3/3) the variation of $K$ with specimen dimensions is often seen as a polynomial series, e.g. for a three-point bend specimen:

$$K = \frac{3QL}{BW^{\frac{3}{2}}} \left[ 1 \cdot 93 \left(\frac{a}{W}\right)^{\frac{1}{2}} - 3 \cdot 07 \left(\frac{a}{W}\right)^{\frac{3}{2}} + 14 \cdot 53 \left(\frac{a}{W}\right)^{\frac{5}{2}} \right.$$
$$\left. - 25 \cdot 11 \left(\frac{a}{W}\right)^{\frac{7}{2}} + 25 \cdot 80 \left(\frac{a}{W}\right)^{\frac{9}{2}} \right] \quad (7)$$

where $a$ is crack length, $W$ is testpiece width, $B$ is testpiece thickness, $Q$ is the applied load, and $L$, the moment arm, is equal to $2W$. Such expressions for $K$ are not capable of direct physical interpretation, because they represent the effect on the *first term* in the series expansion of a general stress distribution of necessarily rather small changes in $a/W$ and the polynomials themselves are 'best-fit' relationships over rather limited ranges. In standard CTS and SEN bend testpieces, $a/W$ values must lie in the range 0·45–0·55 to conform to these relationships.

There are errors in characterizing the stress *at a finite distance ahead* of a crack tip by a single value of $K$ and these errors increase with distance. Moreover, the errors are different in testpieces of different geometry as shown in Fig. 5 (due to Wilson (3/4)). A simple approach to quasi-elastic fracture suggests that an elastic stress analysis can be used to characterize fracture through a value of $K_{crit}$, if $K$ is calculated from the applied stress and a 'notional' crack length modified to include a fraction of the extent of yielding ahead of the crack. For the infinite body in plane stress, the modified stress intensity is given by:

$$K' = \sigma_{app} \sqrt{\{\pi(a + r_Y)\}} \quad (8)$$

where $r_Y$ is the plastic zone radius, given by

---

* See footnote to General notation on page vii.

$$r_Y = \frac{K^2}{2\pi\sigma_Y^2} \tag{9}$$

where $\sigma_Y$ is the uniaxial yield stress. In plane strain, a 'plane strain plastic zone radius', $r_{1Y}$ is taken as $\frac{1}{3}r_Y$ roughly to allow for the effect of constraint in restricting plastic flow. Obviously, the more ductile a material is, the larger $r_Y$ or $r_{1Y}$ will be. From equation (8) this means that a larger correction must be made to the stress intensity, but Fig. 5 shows that this means additionally that the description of stress distribution by $K$-values becomes less and less precise.

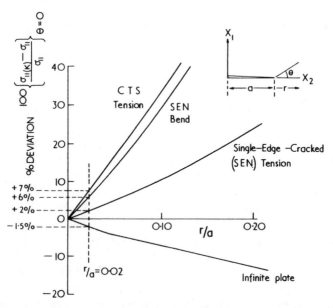

Fig. 5 *Variation between crack tip stress as calculated from K (denoted by $\sigma_{11(K)}$) and that calculated from the full series for $\sigma_{11}$, expressed as the percentage*

$$100\left\{\frac{\sigma_{11(K)} - \sigma_{11}}{\sigma_{11}}\right\} \theta = 0$$

*and plotted vs non-dimensional distance ahead of crack tip $r/a$ [after (3/4)]*

Standard specimen size requirements demand that $r_{1Y}$ should be less than $0.02a$ to allow elastic analyses to be used. For $0.45 < a/W < 0.55$, this also implies that $r_Y \simeq 0.02(W - a)$. Figure 5 shows that, at this limit, the error in using $K$ to characterize stresses in CTS or SEN bend test-pieces is 6–7 per cent positive: in an infinite plate it is 1·5 per cent negative. If the infinite plate were taken to represent a large structure, the use of a $K_{crit}$ value measured on a testpiece to calculate the failure stress in the structure would be subject to an inherent error of some 8 per cent.

## 3.3 ELASTIC/PLASTIC STRESS STATES

A straightforward method of calculating the stress distributon in an infinite body of elastic/non-hardening material containing a central crack and loaded in plane stress tension derives from the work of Dugdale (3/5). Infinities in stresses are removed by loading the crack + plastic zone additionally by point loads of yield load magnitude over the length of the zone. Similar results are obtained by modelling the crack and zone as a continuous distribution of dislocations (3/6). A crack of half-length $a$ is in equilibrium with a yield zone of total length, $d_Y$ under a stress, $\sigma_{app}$, when

$$\frac{a}{a + d_Y} = \cos\left(\frac{\pi \sigma_{app}}{2 \sigma_Y}\right) \tag{10}$$

As $\sigma_{app} \to \sigma_Y$, so $d_Y \to \infty$, and a state of *general yield* is obtained: as $\sigma_{app}/\sigma_Y \to 0$, so

$$d_Y = \frac{\pi^2}{8} \cdot \frac{\sigma^2_{app} a}{\sigma_Y^2} = \frac{\pi}{8} \cdot \frac{K^2}{\sigma_Y^2} \tag{11}$$

which is approximately twice the plastic zone 'radius', $r_Y$ (see equation (9)). A further feature of the yielded crack is that there is a discrete opening at the crack tip which does not occur for an elastic crack (see also (3/7)). This opening: the *crack opening displacement* (COD) is given by:

$$\delta = \frac{8}{\pi} \cdot \frac{\sigma_Y}{E} a \ln \sec\left(\frac{\pi \sigma_{app}}{2 \sigma_Y}\right) \tag{12}$$

An alternative fracture criterion can be postulated if it is supposed that fracture occurs at a critical COD, $\delta_{crit}$. The fracture stress, $\sigma_Y$, for a crack length, $a$, would then be the value of $\sigma_{app}$ required to produce the critical displacement. The argument is appealing if equation (12) is written in the form, for small values of $\sigma_{app}/\sigma_Y$ (i.e. quasi-elastic fractures):

$$\sigma_F = \sqrt{\left(\frac{E \sigma_Y \delta_{crit}}{\pi a}\right)} \text{ or } \delta_{crit} = \frac{K^2_{crit}}{\sigma_Y E} \tag{13}$$

which is equivalent to equation (4) if $G_{crit} \equiv \sigma_Y \delta_{crit}$. This identity is appealing from the virtual work method of calculating $G$: the incremental work done per unit thickness for a crack advance $\delta a$ is achieved by a force per unit thickness $\delta_Y \delta a$ moving through a displacement $\delta_{crit}$. The instability aspect of the $\delta_{crit}$ criterion is, however, unclear. A critical value can be specified to characterize crack extension, but this extension does not necessarily occur in an unstable manner. What drives a crack is the release of *elastic* energy, not the accumulation of plastic work. It would be reasonable to expect that the release of elastic energy would be greater as $\delta_{crit}$ increased, since $\sigma_{app}$ is increased, but this does not seem to be the case in plane stress for the yielded crack in the infinite body: the larger value of $G$ is expended completely, simply in producing a larger plastic zone. The energetics of similar increases of applied stress for elastic/plastic cracks in finite bodies do not seem to have been treated in such detail.

In plane strain, the elastic/plastic stress state is rather too difficult to treat analytically, but finite element results obtained by Rice (3/8), using special crack-tip elements to allow for the angular distribution of shear strain at the crack tip, produce forms similar to equations (11) and (13).

For small-scale yielding, the maximum extent of the plastic zone, $R_Y$, is given by:

$$R_Y = 0\cdot 155 \frac{K^2}{\sigma_Y^2} \tag{14}$$

cf. equation (11) and $\delta$ is given by

$$\delta = 0\cdot 49 \frac{K^2}{\sigma_Y E} \tag{15}$$

cf. equation (13).

The effect of finite boundaries on such results has been studied recently by Carlsson and Larsson (3/9), for standard testpieces at the ASTM or BSI limit (the nominal '$r_{IY}$' $= 0\cdot 02a$, following equation (9)). They find that the stress state is not characterized simply by a single $K$ value and that this affects the size of zone. The main effect is not that of the deviation of the stress across the cracking plane (Fig. 5) but of the degree of biaxility induced by the principal stress acting collinear with the crack. This is characterized by two parameters: the $K$ value and a constant '$T$'-stress. The effects on zone size in testpiece geometries other than four-point SEN bend is quite large, but parameters such as COD show less variation (see also (3/10)).

Effects of $(W - a)$ on fracture toughness have been less well explored, but recent experimental results, on otherwise 'standard' $K_{IC}$ tests, suggest that $a/W$ values up to 0·8 may be used in four-point SEN bend, without significantly affecting the measured toughness value (3/11). If the zone size at fracture is small, the far boundary has to be brought quite close to the crack tip before its effect on spread of plasticity is noticeable. It may be important that, in the geometry studied, the $T$-effect is small.

Attempts to calculate fracture stresses for bodies in planar stress states, when the plasticity preceding fracture is less than that corresponding to general yield, but still not small, have followed two main courses, setting aside the work of Hayes and of Turner, who write elsewhere in this book. One involves the use of a COD approach: the other is based on the $J$ integral (3/12). Although the use of $\delta_{crit}$ is not explicit (indeed, denied) in the work of Heald, Spink and Worthington (3/13) their approach to high stress fracture can be readily appreciated by substituting (equation (13)) $K^2/_{crit}$ $\sigma_Y E$ for $\delta_{crit}$ (equation (12)) with $\sigma_{app} = \sigma_F$ to derive a relationship between the apparent toughness, $K' = \sigma_F W \pi a$, and the true value, $K_{crit}$. Substitution of the UTS, for $\sigma_Y$, as a flow stress, enables results to be treated up to, and even above, general yield. Neglect of the effects of finite boundaries appears recently to have been remedied by Chell (3/14) and Kirby.

The path independent $J$-integral, or energy-momentum tensor, characterizes, for non-linear elastic deformation, the total potential energy release rate, as a crack is extended, around any chosen contour, taking into account not only changes in stored energy density but also any work done by tractions acting on the contour. Recently, attempts have been made to use the $J$-integral to calculate energy release in the elastic/plastic situation. For small-scale yielding, the plastic zone behaves rather as a non-linear-elastic enclave if $J$ is evaluated on a contour that does not thread the zone and, in the limit, $J \to G$ as the non-linear region shinks to zero. Its use to characterize high stress fracture by means of a critical $J_{crit}$ value seems to be fraught with hazard, since the energy release from a plastic region is totally dissimilar to that from a non-linear-elastic region at the same strain level, yet good experimental confirmation of the $J_{crit}$ criterion is maintained (3/15).

At this level neither approach does other than take a fracture stress in a mixed elastic/plastic situation and try to calculate from this an equivalent critical elastic energy release rate which may then be shown in other geometries to characterize fracture. There are techniques available to measure COD and $J$ in small, fully yielded testpieces, to try to obtain values that may be of use in predicting the failure stresses of large structures, but such values pertain to the onset of crack extension, which may or may not give instability in the larger piece. The micromechanisms of the fracture are of importance here, but, before discussing these, it is of interest first to consider macroscopic aspects in relation to testpiece thickness.

## 3.4 TESTPIECE THICKNESS

Until now, discussion has centred on the ideal states of plane stress and plane strain. Differences in elastic/plastic deformation have been summarized, but effects of triaxial constraint have not been considered explicitly. These should now be emphasized, with reference to Fig. 6, which shows the effect of testpiece thickness on the measured value of $G_{\text{crit}}$ (3/16). In region $A$, the through-thickness stress is small; yielding and flow eventually concentrate onto 45° planes and (slant) fracture occurs by a sliding-off mechanism. In region $C$, the proportion of shear lip (plane stress region) is insignificant with respect to the thick (plane strain) central region and fracture occurs in a manner controlled by the plane strain (square) fracture mechanism. This is brittle in these aluminium alloys, which fracture by fine-scale void coalescence, and would be likely to be so also for cleavage or intergranular fracture modes in steel. From results such as those in Fig. 6, the standards specify $B > 2 \cdot 5 \; (K_{\text{IC}}/\sigma_Y)^2$ to ensure that the toughness does correspond to 'plane strain'. The mere attainment of 'plane strain', however, does not guarantee a brittle fracture. The plastic zone size preceding fracture must be small also with respect to crack length and ligament, so that the crack tip strain associated with fracture is of major importance.

In region $B$, a specimen behaves rather like a laminate, in which both square (plane strain) and slant (plane stress) modes give significant contributions to the toughness. Typically, a 'thumbnail' of square fracture is produced at a low load and tunnels into the centre of the specimen, being restrained from catastrophic propagation by the need to continue to deform the plane stress edge regions, which cannot fracture at such a low load (or displacement). If it extends suddenly, there may be a marked 'pop-in' on the load/displacement trace, but it may grow slowly, giving rise to greater curvature in the trace than plasticity alone would produce. The rationale of

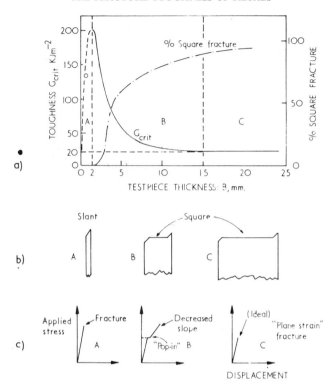

(a) Variation of toughness, $G_{crit}$, with thickness, $B$, for Al Zn Mg alloy.
(b) Schematic fracture profiles.
(c) Schematic stress/displacement curves [after (3/16 and 3/2)].

Fig. 6

the *offset procedure* specified in the $K_{IC}$ standard (3/3) is that it provides some means of deciding whether an amount of crack growth equal to the plastic zone size has occurred at the interception point, $Q_5$.

An underestimate of the total toughness of the 'laminate' testpiece in region $B$ can be made (3/2) by summing contributions from the central and edge regions, but for a piece containing a growing thumbnail, the failure point is not easy to deduce. As the thumbnail grows under increasing load, $G$ increases markedly, but total instability does not occur, because the edge regions still require increasing work. An '*R-curve*' is produced in which $G$ is plotted against increase in crack length, $\delta a$. There is some evidence that instability occurs at a constant *absolute* increase, when, presumably, a geometrical through-thickness condition is met.

We now examine micromechanisms of square fractures, associated with plane strain deformation, if not with quasi-elastic instability.

## 3.5 MICRO-MECHANISMS OF FRACTURE

It proves convenient to divide types of fracture into two categories: *cracking* processes ad *rupture* processes. The former is ideally modelled by cleavage fracture in mild steel: the latter by ductile void coalescence.

Cleavage fracture in mild steel depends on two components: the initiation and the propagation of microcracks, usually formed in grain boundary carbides. At moderately low temperatures, microcracks are initiated by slip dislocations and the fracture criterion for a yielded region then becomes simply the attainment of sufficient local tensile stress, $\sigma_F$, to propagate the microcracks through the ferrite grains. The stress can be measured conveniently in notched bars and is found to be substantially independent of temperature (**1/17**).

A value of $K_{IC}$ for cleavage fracture may be deduced, if it is assumed that the critical $\sigma_F$ value must be attained at some *fixed distance* ahead of a crack tip, i.e. at a position where a suitable (cracked carbide) nucleus is to be found. Then, if the magnitude of the uniaxial yield stress is known, the size of the plastic zone (and hence the $K$ value) required to elevate the tensile stress at the critical position to a sufficient value may be deduced from Tracey's finite element results (**3/18**), Fig. 7. Taking the critical distance as one or two grain diameters, Ritchie, Rice and Knott (**3/19**) calculated values of $K_{IC}$ as a function of temperature for a high-nitrogen mild steel. These calculated values are compared with experimental values in Fig. 8. It is clear that the model has substance, although the excellent agreement found for a distance of two grain diameters should not be regarded as other than fortuitous. A correct interpretation depends on the statistical distribution of carbide nuclei (**3/20**).

The model has been used with success to calculate $K_{IC}$ for an A533B pressure vessel steel from results on small specimens (**3/21**) and has also been applied for intergranular fractures (**3/22**). The problem with small specimens is that triaxial stresses are relaxed by gross yielding at relatively low temperatures, so that the transitions from brittle to ductile behaviour in small and large specimens are not identical. In general, the results appear

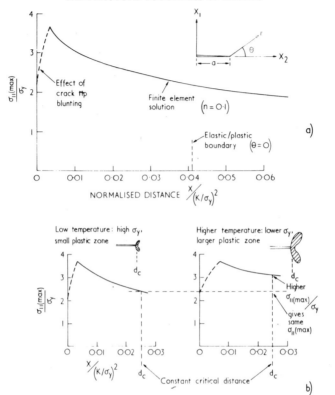

(a) Distribution of stress $\sigma_{11}$ ahead of crack tip under small-scale yielding (after (3/18)).
(b) Schematic explanation of Ritchie, Rice and Knott (3/19) model. A larger plastic zone is needed at high temperatures to obtain the same tensile stress.

*Fig. 7*

to be successful because the propagation of a microcrack across a ferrite grain leads almost instantaneously to total instability. This situation would not hold in steels where extensive single-grain microcracks were formed under increasing load.

When the fracture is produced by fibrous mechanisms, the difference between the first crack extension and instability is increased. In a moderately ductile material, possessing high work-hardening capacity, the crack tip blunts to a rounded shape and coalesces with a void formed around the nearest inclusions to the tip by internal necking, Fig. 9 (3/23). The clean

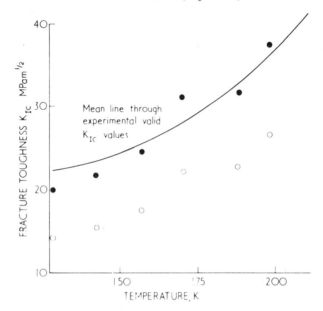

*Fig. 8 Predictions of Ritchie, Rice and Knott (**3/19**) model compared with experimental $K_{IC}$ results.*

surface running up to the first cusp is a 'stretch zone', of width approximately $\delta_i/\sqrt{2}$ (**3/24, 3/25**). Rice and Johnson (**3/26**) have calculated the crack tip opening for hardening materials and Fig. 10 shows $\delta/X_0$ plotted vs $X_0/R_0$, where $X_0$ and $R_0$ are the inclusion spacing and radius respectively. There is obviously excellent agreement between the predictions and the experimental values (**3/23**) for the COD at initiation, $\delta_i$ (see Fig. 6). Only if instability is preceded by an infinitesimal amount of fibrous growth, can a $\delta_i$ value be related to a critical value of $K$, via equation (15). However, an experimental result for $\delta_i$ in A533B, 0·175 mm, gives $K_{IC} = 190$ M Pa m$^{\frac{1}{2}}$, which agrees well with full-scale experimental values at 293 $K$, where the amount of fibrous thumbnail was vanishingly small (**3/21**).

The blunting model sets a lower limit to $\delta_i/X_0$ of 0·5, when the slip-line from a rounded tip first envelops a void ahead of it. In high strength

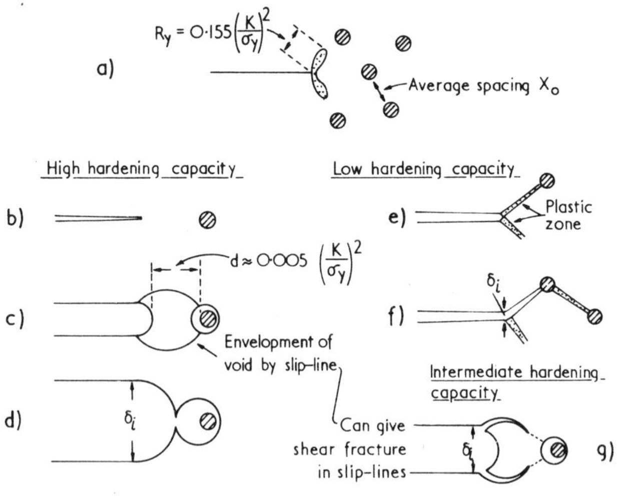

(a) general picture of crack tip and inclusions.
(b), (c), (d) internal necking (after (3/26)).
(e), (f) shear decohesion (after (3/27) (3/28)).
(g) termination of internal necking by shear decohesion (after (3/28)).

*Fig. 9  Modes of ductile fracture separation.*

material and prestrained material it is, however, possible to obtain shear decohesion along intense slip-bands emanating from a sharp crack ($\delta \ll X_0$). These progress from inclusion to inclusion, giving a 'zig-zag' appearance to the fracture, see Fig. 9, but the critical event is the nucleation of voids around matrix carbides (**3/27, 3/28**). To permit slip-bands to remain intensely localized, the hardening capacity of the matrix must be small. This factor appears to be of importance in very high strength steels where $K_{IC}$ decreases with increasing yield strength, even though the micromechanisms of fracture remain fibrous. The only way in which $\delta$ can be decreased in fibrous fracture is to concentrate strain more effectively through reduced hardening capacity. The situation in steels is unlike that in some aluminium alloys, where both $K_{IC}$ and $\sigma_Y$ may increase with decreasing temperature. Following Hahn and Rosenfield (**3/29**) an approximate linear relationship may be obtained between $K_{IC}$ and $n\sqrt{\sigma_Y}$, or between $\delta_i$ and $n^2$, in such alloys, where $n$ is the work-hardening exponent (**3/30**).

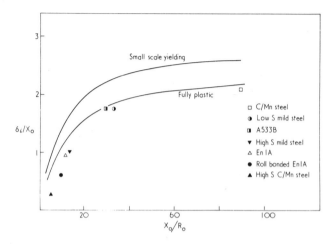

*Fig. 10* Predictions of Rice and Johnson (**3/26**) model with experimental results (after (**3/23**)). The COD at initiation, $\delta_1$, is plotted, as a fraction of inclusion spacing, $X_o$, vs. the ratio of inclusion spacing to radius $X_o/R_o$.

## 3.6 CONCLUSIONS

Fracture toughness was introduced as the link to relate fracture stress to defect size. Given the inherent errors in stress analysis of structures (confused by residual stresses and so forth) and in sizing of defects by ultrasonic techniques, the standard $K_{IC}$ test, giving figures reproducible to $\pm 10$ per cent, is quite sufficient for those applications where 'valid' results can be used. The limiting plastic zone size for brittle fractures in steel can be related to the micromechanisms of fracture, at least in simple systems. The mechanisms of unstable void linkage and the role of hardening capacity in high strength alloys need further study.

For more ductile materials, the COD and $J$-integral approaches try to characterize fracture in ways which are open to discussion, but which can be tested by experiment. To speculate on future developments, it is likely that behaviour in small and large testpieces will be able to be reconciled through the use of COD or $J$ or both. In design, however, the situation is different. At present, the use of COD in design is achieved by experimental strain measurements in large plates and estimates of strain levels around defects in structures (3/31).

Calculations of COD or $J$ can be made using finite element techniques, but part of the assumption is that, if failure is controlled by fracture in a testpiece, it is controlled by a similar type of fracture in the structure. Many testpieces, containing long cracks (high $a/W$) do becomes unstable as a result of fast fracture, but many structures, particularly those made in ductile structural steel and containing short defects (low $a/W$), are likely to fail by general plastic collapse at a stress lower than that needed to cause propagation in a fibrous fracture mode. A design curve which incorporates the transition from fracture to collapse must then be devised and the establishment of such curves for ductile steels presents a major challenge for the future, compounded by the use in service of relatively thin sections, in which both slant and square fracture modes can occur.

# 4
# Yielding Fracture Mechanics
## C. E. TURNER
*Mechanical Engineering Department, Imperial College*

Yielding fracture mechanics seeks to find a relationship between applied stress, crack size and material toughness that is independent of the geometry of a component when fracture occurs after significant degree of yielding. The crack opening displacement, $\delta$, and the $J$ contour integral are two proposals for describing the stresses and deformation at the tip of a sharp crack embedded in a region of yielding material. The concepts can be related in the form $J = M\,\sigma_Y \delta$ where $\sigma_Y$ is the uniaxial yield stress, and $M$ a factor with a value between about 1 and 2·5. The concepts are still under development. Either term can be chosen as a measure of the severity of crack tip deformation in a given material with the onset of crack growth in monotonic loading occurring at a critical value, $\delta_c$ or $J_c$, for a given thickness. Experimental evidence so far is in broad support of this picture but there remains uncertainty over the degree to which $\delta_c$ or $J_c$ is independent of geometry and the extent to which stable crack growth prevents the usage of one simple criterion of fracture for all structural configurations.

**Notation addition to General notation on page vii**

| | |
|---|---|
| $A$ | constant in a non-linear stress-strain law |
| $c$ | effective crack length in the Dugdale model |
| $e$ | nominal strain in the absence of a crack |
| $e_{ij}$ | generalized components of strain |
| $e_p$ | plastic component of strain |
| $e_Y$ | uniaxial yield strain $\sigma_Y/E$ |
| $E'$ | reduced modulus: $E' = E$ for plane stress; $E' = E/(1 - v^2)$ for plane strain |
| $J$ | contour integral defined equation (10) |
| $N$ | index of stress for work hardening stress-strain relationship |

*The MS of this paper was received at the Institution of Mechanical Engineers on 28 May 1975 and accepted for publication on 22 August 1975.*

$Q$ load
$Q_L$ limit load
$r, \theta$ polar co-ordinates, origin at the crack tip
$s$ arc length
$t$ crack face restraining stress in the Dugdale model
$T_i$ generalized component of forces acting on a stated surface
$u_{ij}$ generalized displacements in the co-ordinate directions
$x, y$ catesian co-ordinates
$w$ strain energy density
$\sigma$ nominal stress on the crack plane in the absence of a crack
$\sigma_u$ ultimate tensile strength
$\sigma_{ij}$ generalized components of stress
$\delta_i$ crack opening displacement at which the crack first grows
$\omega$ work done under the load deflection curve

## 4.1 INTRODUCTION

The prime objective of yielding fracture mechanics is to described the fracture circumstances for a material of limited ductility in the presence of a defect. It has not been employed for the description of completely ductile fractures following large deformation and flow comparable to that found in a conventional tensile test of a ductile metal. In short, the problem is 'brittle fracture' where brittle means 'less than the anticipated ductility' rather than the linear elastic event implied by the strict meaning of the word 'brittle'. Nothing is said about the reasons on the microstructural scale for how or why the brittle fracture occurs. As with other branches of fracture mechanics, a means is sought whereby fracture observed in the laboratory can be interpreted in terms of a material property, toughness, and mechanical features such as stress level or geometry of component and then applied to avoid fracture in real structures. Just as in linear elastic fracture mechanics (lefm) one approach is to characterize the singular stress, strain and displacement fields around the tip of a sharp crack by a single parameter (**4/1**) so too in yielding fracture mechanics efforts have been made to find a one-parameter characterization of the elastic-plastic stress and strain fields local to the crack tip. It is argued, as in lefm, that if such a parameter can be found, then for a given composition, temperature, strain rate and environment a process such as brittle fracture, supposed to be governed by circumstances at the crack tip, must occur when this parameter reaches a critical value. As in lefm, the argument is presented for an idealized two-dimensional case of plane stress or plane strain, since it is readily demonstrated that fracture of otherwise similar thin or thick pieces occurs with differing values of toughness. It has also been argued (4/2) that variations of in-plane geometry will cause different degrees of biaxiality so that a unique value of one fracture parameter will not occur for differing circumstances. The problem thus resolves itself into two parts: can a simplified model of elastic-plastic crack tip behaviour be established such that the dominant stresses and strain close to the crack tip are simply described: if such a model can be established will it in practice be suffiiciently invariant to conditions remote from the crack—notably crack length, overall geometry and applied stress systems—to serve a useful purpose, i.e. to allow a laboratory test to be used to predict and thus avoid fracture in the real structure?

## 4.2 SLIP LINE FIELDS

Complete analytical solutions to the elastic-plastic crack tip stress fields using a yield criterion and the conventional incremental stress strain laws of plasticity are not known. The simplified model first used for a number of plasticity problems where deformations were large was the rigid-plastic slip-line field analysis (4/3). This model neglects elasticity and work hardening and was not directed towards fracture problems. It assumes the plastically deformed material moves in rigid 'blocks' so shaped by the loading system as to slide over one another along lines of maximum shear stress. The method serves as a first pointer in fracture problems, showing for example that triaxial stresses are likely to build up where the slip line pattern as traced from a free boundary is curved, and that negligible triaxiality is induced, when the slip line pattern is straight (Fig. 11).

The maximum load a member can carry, the limit load, is predicted reasonably well. Since only extensive plasticity is modelled and the possibility of a fracture event intervening is not considered, slip line field solutions provide no measure of the deformation a component can sustain for a given degree of ductility of the material. The method is however used to guard against failure by plastic collapse in either notched or unnotched bodies, i.e. failure by overload of structure made of a material of 'unlimited' ductility.

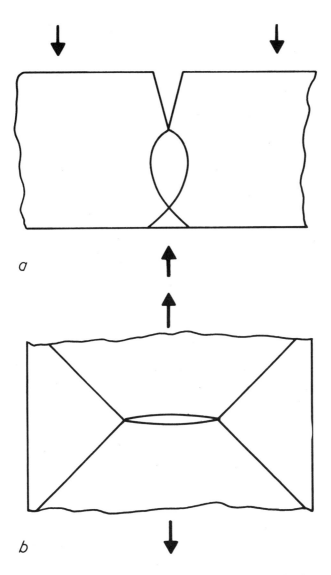

*Fig. 11* Typical slip line field solutions showing (a) curved slip lines with high triaxial stresses induced at the notch root in bending, (b) straight slip lines with no elevation of stress at the crack tip for centre cracked plate.

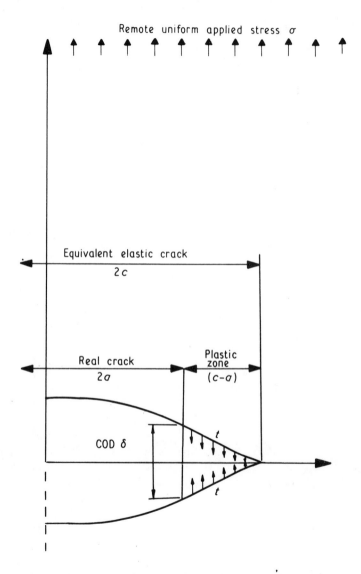

*Fig. 12  The Dugdale 'strip-yield' model for crack tip plasticity.*

## 4.3 CRACK OPENING DISPLACEMENT, COD

Another model that is the basis of much of the present generation of fracture work, is based on linear elasticity, neglecting plasticity. In one sense this model is even cruder than the slip line field model but by retaining elastic deformations and ignoring the effects of extensive plasticity a relevance to fractures after restricted amounts of plastic flow emerges. The model, proposed separately by Dugdale (4/4) and Barenblatt (4/5), considers an infinite plate with a central crack, length $2a$, subjected to a remotely applied uniform stress $\sigma$. The plasticity at the crack tip is represented by a notional increase in the crack length to some value, $2c$, with the faces of the 'crack' over the distances $(c - a)$ from both ends partly restricted from opening by a restraining stress, $t$, acting directly on the 'crack' faces, Fig. 12. This model, often called the 'strip yield model' was interpreted in terms of fracture by Wells (4/6). Wells had noticed that in practice the tip of a slot subjected to plastic deformation opened with a near square ended contour, giving a definite tip opening—the crack opening displacement—(COD) as Fig. 13. He proposed that the COD, $\delta$, was a measure of crack tip deformation and that fracture might occur when a critical value of this parameter, $\delta_c$ was reached. This proposal was pursued experimentally and theoretically by Burdekin and Stone (4/7) who showed that it was broadly consistent with fracture results from large tension and bending tests, although agreement at a more detailed level was by no means complete.

At about the same time as these developments occurred a similar model was proposed by Bilby et al (4/8) using dislocation theory. It was clear that their model represented plane stress and might thus be more relevant to thin section behaviour than to thick section. Metallurgical studies by etching (4/9) also showed that for thin sections a line of plasticity did indeed develop ahead of the crack tip very much as pictured in the Dugdale or Bilby model, though at 45° oblique through the thickness rather than normal to the plate surface. For thick sections, the clear implication of the slip line field model is that a degree of constraint develops at the crack tip with triaxial stresses, all tensile, such that the restraining stress, $t$, would be some value such as $M\sigma_Y$, and from slip line field theory $1\cdot15 < M < 2\cdot98$.

However, in early applications of this model (4/7), $t$ was equated to the uniaxial yield stress, $\sigma_Y$, and relationships obtained for the length, $2c$, to

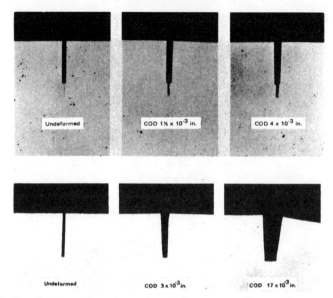

*Fig. 13  Development of crack-opening displacement at the tip of saw-cut.*

Fig. 13 is reproduced by kind permission of the Welding Institute.

which the notional plastic zone extended, in terms of the real length, $2a$, and the applied stress, $\sigma$,

$$a/c = \cos \frac{\pi \sigma}{2\sigma_Y} \tag{1}$$

The COD, $\delta$, at the tip of the real crack was then evaluated as

$$\delta = \frac{8\sigma_Y a}{\pi E} \log \sec \frac{\pi \sigma}{2\sigma_Y} \tag{2}$$

By expanding the log sec term it was found that (**4/7**)

$$\delta = \frac{\pi \sigma^2 a}{E \sigma_Y} \left[ 1 + \frac{\pi^2}{24} \left( \frac{\sigma}{\sigma_Y} \right)^2 + - - \right] \tag{3}$$

Noting that in lefm, for the infinite plate with centre crack $2a$,

$$G = K^2/E = \pi \sigma^2 a / E \tag{4}$$

it is at once seen that the first term in equation (3) corresponds to

$$G = \sigma_Y \delta \qquad (5)$$

If the plastic zone correction factor for plane stress is used, so that the effective crack length $2c = 2(a + r_p)$ where $r_p = \dfrac{1}{2\pi}\left(\dfrac{K}{\sigma_Y}\right)^2$, then the modified lefm expression is

$$G = \frac{\pi \sigma^2 a}{E}\left[1 + \tfrac{1}{2}\left(\frac{\sigma}{\sigma_Y}\right)^2\right] \qquad (6)$$

which together with equation (5) differs from equation (3) by a fairly small amount in the coefficient of the second term. The relationship $G = t\delta$ is consistent with the work done to close an element of crack, from the reasoning of mechanics on either the macro or dislocation level, so that this model, used with the restraining stress $t = \sigma_Y$, appears to be a logical extension to lefm for plane stress.

Various attempts were made at a later stage to represent plane strain by:

$$G = M\sigma_Y \delta \qquad (7)$$

Experimental evidence (**4/10**) suggested that $M \simeq 2 \cdot 1$ for compatibility with lefm in cases of limited ductility where either concept could be applied, but the subsequent engineering design application proposals remained based on the Burdekin Stone infinite plate model with $t = \sigma_Y$, i.e. $M = 1$ and were applied to cases of fracture after extensive plasticity rather than to near lefm situations. A modified approach has been suggested by Heald et al (**4/11**) with $t = \sigma_u$ (the ultimate tensile strength) and $\delta$ expressed in terms of $K$, using equation (4) and (5). The model degenerates to lefm for brittle materials and to a failure stress of $\sigma_u$ for ductile ones, and thus matches observed failure stresses, at least in simple tension type problems.

## 4.4 CALCULATION AND MEASUREMENT OF COD

One of the practical drawbacks of the strip yield model has been the absence, until recently, of a significant number of solutions for different geometries (**4/A–4/C**)*. Computational methods are of course possible for strip yield model itself (**4/12**). Elastic-plastic finite element studies can be made from which $\delta$ can be evaluated. In the former $\delta$ is inherent from

---

\* *References (**4/A–4/L**) are included as an addendum to the General Reference Section.*

the formulation. In the latter there is some difficulty in identifying precisely the point in the crack tip region for which δ should be evaluated, since an absolutely 'square tipped' crack is not found in finite element studies using conventional elements. The absence of solutions for even simple test piece geometries greatly hindered the early verification of the concept that $\delta_c$ might be constant at fracture of the same material for different geometrical circumstances since even such solutions as have emerged are of more recent date than the development of COD concepts described here.

In early experimental work, saw cut notches some 0·006 in (0·15 mm) wide were used and the tip COD measured by a transducer with paddle or spade shaped tip lightly spring loaded against the faces of the notch. An extensive test program was conducted under the auspices of the then Navy Department Advisory Committee on Structural Steel, reported in (4/13). The results showed the importance of using a sharp notch, usually fatigue cracked, if minimum toughness values were to be found. Probes inserted in the notch reaching to the crack tip were impracticable for cracks as sharp as those produced by fatigue so that a method was developed of using a clip gauge across the surface (mouth) of the crack from which the tip COD was deduced by a calibration factor (4/13). In (4/13) good agreement was found between COD, δ, as calculated and as measured, and a broad agreement between critical COD at fracture, $\delta_c$, measured in different circumstances as a function of temperature for several structural steels. A great amount of scatter in $\delta_c$ was found in the transition temperature region, attributable in part to the occurrence of slow stable crack growth subsequent to the initiation of crack growth but before complete failure of the test piece. Where this effect was detected and recorded, scatter in test results was appreciably reduced. The problem then arises that if there is not a single unique value of δ identifiable with fracture, what value should be measured? The initiation of growth from the pre-existing crack, occurring at a value, $\delta_i$, is an event of interest. Some value representative of the onset of unstable growth and rapid fracture, or the value of COD at maximum load, $\delta_m$ (which might or might not coincide with onset of unstable growth) may however be more relevant to the final separation process. Subsequent tests (4/14) tend to show that $\delta_i$ is more nearly independent of geometry than $\delta_m$ but the numerical values of δ at crack initiation are small probably implying an unrealistic degree of conservatism in relation to the final fracture event. Other studies report favourably on the validity of the critical COD concept (4/15) but it is difficult to judge whether the several problems mentioned above are there fortuitously minimal, or whether for some reason there is a series of compensating events in the particular tests or materials used. The problem of stable crack growth and its interpretation in fracture problems remains unresolved to this day in connection with both COD and any other elastic plastic fracture model, although some insight is now being gained into the final instability by equating the plastic extension (itself measured in terms of COD for a deep notch problem) to the elastic contraction that would occur remote from the notch as the crack spread and the component unloaded (**5/D**).

## 4.5 THE USAGE OF COD

To allow the application of COD to actual problem solving a complete design method was necessary. Testing procedures were therefore evolved (as described Ref. (**4/16**)) using deep notched three point bend pieces in which the degree of constraint is severe. COD is measured by a surface clip gauge and calibration formula. The implication is that real service conditions are unlikely to be more severe in respect of geometrical constraint. Recommendations are made to cover other points of uncertainty, such as slow crack growth, that are either conservative, or allow for agreement between the parties concerned to suit the problem in hand. A design curve has been established (**4/17**), (**4/18**) relating a non-dimensionalized COD, $\delta/2\pi e_Y a$, to the nominal applied strain $e/e_Y$, with an inbuilt degree of conservatism such that all the test data falls on the safe side of the curve. Clearly the curve will be more conservative for some applications than for others, but that is inevitable with a single design curve approach. The third element contributing to the design usage of COD is an assessment of the significance of defects (**4/19**). Rules are suggested for the treatment of adjacent defects, part through thickness defects or buried defects approaching a free surface, non-planar defects and so on, so that the actual defect or group of defects suspected in a structure can be represented by an equivalent simple defect for interpretation with the design curve.

## 4.6 THE $J$ CONTOUR INTEGRAL

A third model of the elastic-plastic crack tip is based on non-linear elasticity, usually expressed in power law form, giving a single analytical expression over the whole strain range of interest.

$$\frac{e^p}{e_Y} = A \left(\frac{\sigma}{\sigma_Y}\right)^N \tag{8}$$

Hutchison (**4/20**) used this type of formulation, to study the local crack tip stresses for a crack in an infinite plate subjected to uniaxial tension. The plastic strain term, equation (8) was additional to a linear elastic term, for which a Heneky-Mises equivalent stress defined a 'yield' criterion for the end of the linear regime. The region of non linear behaviour was extensive in relation to crack length but remote from the crack the stress remained linearly elastic.

Plastic stress and strain intensity factors were derived at the tip of a crack, showing a dependence on distance, $r$, from the crack tip, of the form

$$\text{stress } \sigma \propto \frac{1}{r^{1/N+1}} \sigma(\theta) \tag{9}$$

$$\text{strain } e^p \propto \frac{1}{r^{N/N+1}} e(\theta)$$

displacements $u \propto r^{1/(N+1)} u(\theta)$

The angular distributions wrt $\theta$ around the line of the crack are given in the original paper but are not directly relevant here. These equations relate to the first term only in a series solution in powers of $r$ for which this first term dominates all others as $r \to 0$. The method thus parallels that used by M. L. Williams in his derivation of the lefm $K$ factor (**4/1**). It will be noted that if $N \to 1$ the linear elastic singularities for stress and strain of power $-\frac{1}{2}$ are recovered; that the displacement term is not singular and that the singularity in energy density formed by the product $\sigma e$ is of the order $-1$, all as expected. Similar results were obtained separately by Rice and Rosengren (**4/21**).

Rice had previously formulated (**4/22**) a term called the $J$ contour integral defined to be

$$J = \int_\Gamma w\, dy - T_i \frac{du_i}{dx}\, ds \tag{10}$$

for a crack aligned in the $x$ direction. Here $\Gamma$ is any contour from the lower crack face anticlockwise around the crack tip to the upper face. $s$ is path length along this contour, $w$ is the strain energy density of the form $\int_0^e \sigma_{ii} de_{ii}$ and $T_i du_i$ are work terms when components of surface tractions on the contour path, $T_i$, move through displacements, $du_i$. The integral was shown to be independent of choice of path for a crack with stress free faces. For the non-linear elastic material being discussed this term was shown to equal the potential energy release rate with crack advance (on a unit thickness basis)

$$J = -dP/da \tag{11}$$

It is thus conceptually equal to the Griffith term $G$ (4/23) except that $G$ implies a linear elastic event whereas $J$ implies a non-linear elastic event. The contour integral had previously been developed separately by Eshelby (4/24) and Cherypanov (4/25) but its current usage in fracture work stems from the integration of the Rice formulation into the Hutchinson/Rice and Rosengren stress analysis solutions. Hutchinson (4/20) gave the proportionality factor for equation (9) in terms of plastic intensity factors and showed that for the so-called 'small scale yielding' situation when linear elasticity still dominates the regions remote from the crack tip, these plastic intensity factors could be expressed in terms of $J$ and some tabulated constants $I$ the value of which depended on $N$. The whole constant term degenerated to one proportional to $K^2$ for small scale yielding. The drawing together of these threads into a specific formula was given by McClintock (4/26). The exact form depends on the statement of the stress strain law and the inclusion or not of constants or non-dimensionalising factors which differ from reference to reference. For the stress–strain law as stated equation (8) the proportionality constant in equation (9) is

$$\left(\frac{JE}{\sigma_Y^2 IA}\right)^{1/(N+1)} \tag{12}$$

Thus for any given hardening law $N$, modulus $E$ and limit of linearity $\sigma_Y$ the crack tip stresses are a function of $J$. It is then argued that events such as fracture, controlled by conditions at the crack tip, must be describable in terms of $J$, just as in lefm they are describable in terms of $K$.

The use of $J$ as a crack tip parameter in *non-linear elasticity* is thus established on both energetic [equation (11)] and characterizing grounds [equation (12)] in close parallel to the position in lefm (4/1). The clarity of the argument is however lost in *plasticity*. The non-linear but still elastic material is identical during loading to a material following 'total' or 'deformation' plasticity (4/3) in which the stress strain property arrived at is a function only of the final loading point considered and not of the stress–strain path by which the final load has been arrived at. Real materials are generally accepted as obeying 'incremental' laws under which the final state *does* depend on the loading history. For monotonic loading in simple tension the stress–strain history of any element of material, such as the ones near the crack tip, probably differs but little according to the theory

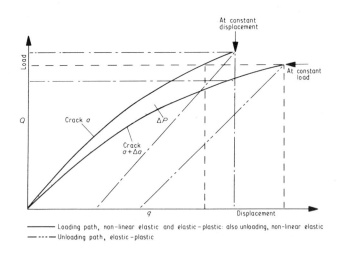

*Fig. 14  Evaluation of J from areas under the load-deformation diagram.*

adopted. If obvious changes in stress ratio occurred by adopting some complex loading history, or if unloading occurs, the two theories differ, possibly by large amounts. It is thus argued that the so-called HRR solutions (**4/20, 4/21**) are representative not only of non-linear elastic behaviour, but plasticity (including 'incremental' plasticity if only simple monotonic loading occurs) with no unloading: i.e. up to onset of crack advance (at which point some elements of material must be unloaded) the crack tip stress and strain fields are described by $J$ using [equation (9) and (12)] to within an acceptable degree of approximation. It must be noted that $J$ here means the contour integral equation (10) shown to be path independent for non-linear elasticity. The physical meaning of $J$ as an energy release rate equation (11) is lost since in either 'deformation' or 'incremental' plasticity the energy represented by $dP/da$ is no longer potentially available for propagating the crack but has been dissipated in the plastic deformation during loading, except for a remaining linear elastic term (Fig. 14). As a numerical quantity, the value of $dP/da$ and hence the term $J$ can be found from

$$J = -d\omega/Bda \tag{13}$$

where $\omega$ is work done under the load deflection curve on loading plastically deforming pieces each with a successively longer initial crack, but, it is repeated, this term is a work *absorption* rate and not the energy *release* rate available to propagate the crack. The satisfying energy balance concept of Griffith (**4/23**) is lost because there are now three energy terms—potential energy (or strain energy if the fixed deflection case is taken); fracture surface energy (including any local plastic events at the surface of separation as in the Irwin/Orowan modification of Griffith) and also work absorbed in extensive plastic flow. Unless these latter two dissipative terms can be shown to be related in some known way, there seems no means whereby the release of elastic energy, even if evaluated, could be used to predict fracture.

The weaknesses of the development of $J$ as a parameter characterizing the singular near tip stress and strain field around any elastic–plastic crack are primarily the restrictions to (i) no unloading (ii) uniaxial remotely applied stresses (iii) a remote stress field that is still elastic (iv) 'deformation' rather than 'incremental' plasticity. The first restriction has already been commented on, namely, that $J$, if relevant at all, must relate to the onset of crack advance under monotonic loading and not to some later stage of unstable fracture subsequent to slow crack growth. The second and third restrictions appear to be very important and have not to the writer's knowledge been firmly reiterated or disproven. It is not clear whether in the original derivations of the plastic crack tip intensity factors these restrictions were merely arbitrary, vitally necessary or perhaps conditional upon other unspecified aspects of the problem. It is clear that in purely elastic behaviour biaxiality has no effect on $K$. It is clear that with plasticity at the crack tip the extent of plastic zone and its characteristics at the elastic–plastic interface *do* depend on biaxiality (i.e. the superposition of a stress parallel to the crack path) (**4/27, 4/1**). It is not clear whether right at the crack tip this effect is 'filtered out' and that $J$ is a unique measure of the tip stresses or whether there is a remaining influence of biaxiality. It was shown (**5/E**) that the effect of the different degree of stress parallel to the crack, induced by different geometries, gave rather small but definite differences in the near tip displacements, themselves taken as a yardstick for crack tip severity. An effect of biaxiality would be in accord with slip line field predictions (**4/2**) and recent computational studied (**4/28, 4/29**) also point in this direction. Further studies (**5/F**) suggest that the crack tip deformations are as closely similar for a given value of $J$ (but not for a given value of load) even for remotely applied biaxial stresses as for simple test configurations, if conventional small deformation theory is used. Recent studies using large deformation theory (**5/G**) confirm that close similarity between the crack tip deformations in different geometries exist for limited plasticity, but tend to be gradually lost as plasticity becomes very extensive. The restriction to a remote elastic stress field arises naturally in an infinite plate solution. If plasticity reaches the far boundaries of a finite cracked body there are intuitive reasons why this may affect the whole nature of the plastic zone and hence the crack tip, but again the writer is not aware of a definitive statement one way or the other on this point. Computational results mentioned below tend to support that $J$ is still meaningful even if plasticity has spread to the far boundary. The fourth issue, the use of the more realistic 'incremental' theory of plasticity has been extensively studied by Hayes (**4/30**), Boyle (**4/31**) and Sumpter (**4/32**) by finite element computation. Hayes found that within computational error the $J$ integral remained path independent from the far field to the near tip regime. Accuracy was lost very close to the tip where the constant strain elements used obviously give a poor representation of the crack tip region, however small the elements may be made. The results of Boyle and Sumpter confirmed these findings. These studies also showed that the path independence of $J$ was not altered by passing contours partly through elastic and partly plastic elements, or through regions where plasticity has spread to the far boundaries. Both hardening and non-hardening cases were studied. If these computational demonstrations of path independence are accepted as adequate demonstrations of the characterizing role of $J$ for real materials suffering extensive plasticity then the use of $J$ seems to be on a firm basis except for the uncertain effect of biaxiality.

This rather lengthy treatment has been given to introduce $J$ because of the niceties of the argument. In the strictest terms, $J$ cannot be proven to be relevant to elastic–plastic fracture, nor can COD or any other model yet derived. The question is whether the $J$ model is sufficiently useful to merit study. Firstly do materials fracture at a critical value of $J$, $J_c$? The answer, of course, is not clear cut. Secondly, if the first answer is sufficiently promising to pursue for the time being, does $J$ offer scope for analysis of the effects of extent of yielding, geometrical features, relationship between test variables such as load, deflection, work done, etc? The answer is a clear 'yes', and herein lies the main attraction of $J$ at the present stage of development.

## 4.7 CALCULATION AND MEASUREMENT OF $J$ AND DETERMINATION OF $J_{1c}$

Determination of $J$ as a function of load or displacement for a given geometry and stress strain curve may be by computational, approximate analytical or experimental methods. Computational results are based almost exclusively on two dimensional finite element studies. Typical computed results for test piece configurations are shown Fig. 15 in non-

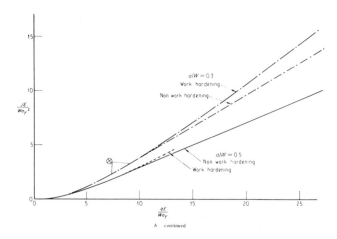

*Fig. 15  Variation of J with (a) load, (b) deflection for selected test piece geometries. For regions × refer to Fig. 16. Note in (a) J is expressed as a non-dimensional coefficient Y\*.*

dimensional form. There is first a region which differs but little from lefm. In this region $J$ increases parabolically with displacement. The load-displacement diagram is nearly linear, and plasticity, though perhaps comparable in extent to the crack length is still restricted to the region around around the crack. Thereafter $J$ increases rapidly with displacement as plasticity spreads over the whole cross section. The load-displacement curve rapidly 'bends over' to approach what for an ideal non-hardening material would be the limit load, Fig. 16.

In the near linear region $J$ can be estimated analytically to an accuracy of a few per cent from

$$J = K^2/E' \tag{14}$$

where $K$ is itself estimated from

$$K = Y\sigma\sqrt{(a + r_p)} \tag{15}$$

$Y$ is here the appropriate geometrical factor in the many known solutions for $K$ and $r_p$ is the plastic zone connection factor (**4/1**). In the second

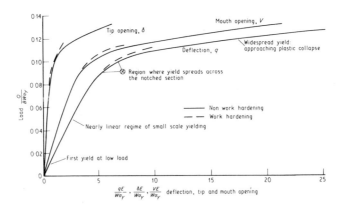

*Fig. 16  Features of load-displacement curves corresponding to development of plastic zones and growth of J. (For region × refer to Fig. 15).*

region where $J$ increases rapidly, the slope of the $J - q$ relationship can be estimated from the limit load if this is known, perhaps from slip line field solutions (4/33). The steps are, neglecting elasticity,

$$\omega = Q_L q \tag{16}$$

where $Q_L$ is the limit load assumed known as a function of geometry. Using equation (13):

$$\frac{dJ}{dq} = -\frac{1}{B}\frac{dQ_L}{da} \tag{17}$$

e.g. in three point bending $Q_L = 1\cdot 5\ \sigma_Y B(W-a)^2/S$ and $\dfrac{dJ}{dq} = 3\cdot 0\ \sigma_Y (W-a)/S$. The elastic based estimate is taken up to the limit load and the constant slope is then grafted on to it to provide a relationship usually in good agreement with computation.

In the first experimental usage of $J$, the term was evaluated as a function of test piece displacement from non-linear load-deflection measurements (Fig. 14) using the relationship $J = -d\omega/Bda$ (4/34). The technique is slow and tedious, but gives a calibration for $J$ in terms of the deflection observed at fracture. Computational studies (**4/31, 4/32**) confirm that the numerical value of $d\omega/da$ agree with the value of the contour integral. More recent work is based largely on the relationship

$$J = 2\omega/B(W - a) \tag{18}$$

that has emerged for deeply cracked three point bend pieces (**4/36**), where $\omega$ is the total work done up to some fixed displacement. This relationship has again been evaluated from computed results and agrees with the value of the contour integral, but equation (18) applies to a very limited range of test piece geometries.

It is beyond the scope of this paper to review the evidence for a specific value of $J$, $J_{1c}$, which can be regarded as a material property characterizing fracture, but brief reference must be made. Early tests using $J$, (**4/34, 4/35**), as with early work on $\delta$, (**4/7**), showed a general relevance of the concept with sufficient merit to warrant further study. Much work then concentrated on simplifying experimental methods for finding $J_{1c}$, as summarized (**4/16**), for conditions of plane strain.

Observation of the work done, equation (18) is the currently preferred method, where value of $J$ at fracture is evaluated from the energy absorbed up to the point of fracture. These further studies revealed the importance of slow crack growth, as previously found in studies of COD so that much current testing is based on use of equation (18) for samples broken at successively earlier stages of test, to detect the onset of cracking.

It remains to be seen whether the concept is tenable for stable growth under elastic–plastic conditions. Such ideas have been used previously in lefm applied to plane stress problems (**4/37**) but have not proved readily applicable since the simplicity of the 'one critical value at fracture' approach is lost, and the fracture event described by a crack growth resistance $R$-curve, in which toughness is a function of the amount of slow crack growth. These concepts have recently been applied to plasticity in terms of a $J$-like term (**5/H, 5/I, 5/J**) using a generalisation of equation (18). The degree of geometry independence remains to be demonstrated, but prediction of unstable crack growth now seems feasible in terms of this resistance curve (**5/K**).

Measurements of toughness in terms of $J$ have so far concentrated on plane strain fracture. To obtain below gross yield fractures and use valid lefm test methods rather large test piece dimensions are required (**4/16**) to maintain a plastic zone very small in relation to thickness and cross section of the test piece. Using yielding mechanics, the extent of plasticity in the plane of the test piece is automatically allowed for by use of $J$, in so far as it has been demonstrated to be valid. It then appears that a far less rigorous thickness criterion is adequate for measurement of a minimum plane stain toughness, $J_{1c}$. The suggested condition (**4/35**) is thickness $B > 25 \, J_{1c}/\sigma_Y$ (or perhaps $50 \, J_{1c}/\sigma_Y$). This would allow pieces some 10 or 20 mm cross section to be used for the determination of toughness of normal structural steels which for valid lefm testing might require sections near 100 mm thick or more. Measurements at mid thickness (**4/38**) tend to confirm that an adequate degree of plane strain is being maintained. This implies that the test conditions required for valid lefm tests relate not so much to thickness per se but to the associated restriction of the extent of in-plane plasticity, a feature no longer necessary when plasticity is accounted for in the description of crack tip stresses and strains. However, this relaxation of restriction on the size of test piece may well be useful only if the micro-mode of separation is the same as in the larger component. There is evidence that if the micro-mode can change, the fracture criteria appears size dependent (**5/L**), although it is not clear that the criterion used (**5/L**) refers to a rigourously defined point of initiation rather than to a subsequent point of complete fracture.

## 4.8 THE RELATIONSHIP BETWEEN $J$ AND COD AND DIFFERENCES IN THEIR USAGE

The relationship between J and COD, whether computed or measured, can be expressed as

$$J = M \sigma_Y \delta \qquad (19)$$

In the Dugdale COD model as currently used, $M \sigma_Y$ equals $t$, the restraining stress at the crack tip. $M$ is then a constant so that in the Dugdale model $J$ and $\delta$ are alternative measures of the same crack tip situation. In more realistic analytical or experimental situations $M$ is not normally found to be constant. As already mentioned, its value lies between 1·15 and 2·98 with many results falling around $2 \pm 0.5$ according to the precise way in which $\delta$ is defined in any analysis other than the Dugdale model. Within that accuracy there is no conflict. The general trend of COD and $J$ with load or deflection is similar. The general scatter of the variation of fracture

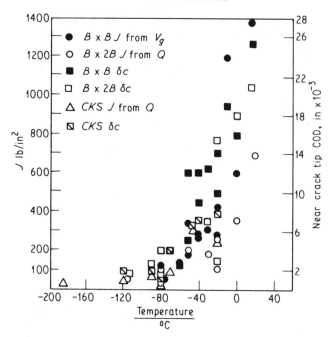

*Fig. 17 Critical COD and J values as a function of temperature for a typical structural steel.*

toughness with temperature (Fig. 17) falls more or less within that uncertainty. Nevertheless the differences between the two concepts could be some twofold and are thus much wider than accepted in lefm.

The mathematical formulation of both linear and yielding fracture mechanics has been reviewed in great detail by Rice **(4/40)**. The current status of COD and $J$ as practical procedures has been reviewed recently **(4/41)**.

Studies of COD made by infiltrating the crack tip **(4/38)** and evaluations of $J$ by equation (18) allow comparison of $\delta_{1c}$ and $J_{1c}$ to be made at the initiation of slow crack growth. First results support $M \simeq 1$ but later results, as yet unpublished **(4/39)** show that on other steels and at greater ductilities $M \simeq 2 \cdot 5$. It thus remains unproven whether there is a unique $J - \delta$ relationship, or whether it depends on both geometry and material.

Another distinction is in the usage that has so far been made of the two concepts. The plane strain values of $J_{1c}$ mentioned above have been used to deduce $G_{1c}$ (whence $K_{1c}$) for use in lefm analyses of components. Most measurements of COD aim to find a critical value, $\delta_c$, representative of the thickness tested, for use in a yielding analysis of a real component. A similar non-plane strain usage of $J$ and associated toughness, $J_c$, could be envisaged but has not so far been employed.

The present status of COD is that despite obvious simplification and weaknesses in the model, the whole gamut of problems from testing to design has been covered by proposals that allow a working usage of yielding fracture mechanics for the assessment of the significance of defects. This overall development is seen by protagonists of the method as more useful to engineers in their current requirements, than concentration on a more rigorous treatment of the fracture criterion, even if a suitable treatment were in fact available.

## 4.9 CONCLUSIONS

In summary it seems that either COD or $J$ offer a reasonable one term description of the conditions at the tip of an elastic–plastic crack. It is likely that neither takes into account the full effects of variation of in-plane constraint, and as with all other fracture mechanics formulations, out-of-plane constraint is known to affect the toughness of materials in a way that so far has to be accounted for experimentally. The COD concept has fairly well developed proposals for overall design usage. The $J$ concept is angled more, at least at the present stage, to the derivation of lefm toughness data from small test pieces that yield before fracture, for use in lefm design circumstances. In broad terms the concepts are not incompatible although differences of interpretation exist in detail.

Whatever its weaknesses, the writer sees no more promising proposals yet on the horizon for the safer design and use of flawed structures than that offered by fracture mechanics. Despite limitations, yielding fracture mechanics is now able to offer a quantitative guide to the acceptance or rejection of flaws and to the assessment of fitness for purpose of defects in engineering components.

# 5
# Evaluation of Stress Intensity Factors

D. J. CARTWRIGHT
*Department of Mechanical Engineering,
University of Southampton*

D. P. ROOKE
*Royal Aircraft Establishment, Farnborough, Hants.*

Some of the more useful methods of evaluating stress intensity factors are presented in a concise form. The stress intensity factor is defined and compared with the more familiar stress concentration factor. The shape factor, the parameter which characterizes the shape of the crack, the orientation of the crack and the proximity of other boundaries, is introduced; the techniques for determining the shape factor are divided into theoretical and experimental. Each method is described with a minimum of mathematical detail; references are given to papers where the methods are more fully described and used to solve specific crack problems. The accuracy and usefulness of the methods is summarized.

## 5.1 INTRODUCTION

A smooth notch in a uniformly stressed elastic solid causes the stress to increase near the tip of the notch. The magnitude of this increase is measured by the stress *concentration* factor $K_t$ for the notch and is the ratio of the maximum stress at the notch tip to the applied stress. As the radius of the notch is reduced to zero, i.e. the notch becomes a crack, the stress at the tip, and hence the stress concentration factor, becomes infinite. Since this applies to all cracks, the stress *concentration* factor cannot be used to distinguish between different cracks.

A parameter which can be used to distinguish between different cracks and which measures the severity of the crack can be obtained from an

*The MS of this paper was received at the Institution of Mechanical Engineers on 28 May 1975 and accepted for publication on 22 August 1975.*

examination of the stress field near the crack tip. The stress $\sigma_{ij}$ at a *small* distance $r$ ahead of the crack tip is given by

$$\sigma_{ij} \simeq \frac{K_N}{\sqrt{(2\pi r)}} \qquad (1)$$

that is, as $r$ approaches zero, the stress near the tip of a crack approaches infinity as $r^{-\frac{1}{2}}$. The constant of proportionality $K_N$, which is different for different cracks, is called the stress *intensity* factor and can be written

$$K_N = Y \sigma \sqrt{(\pi a)} \qquad (2)$$

where $N =$ I, II or III indicates the relative movement of the crack faces (see for example Hayes, chapter 2). $\sigma$ is a stress determined by the loading, $a$ is a crack length and $Y$ a geometrical factor* which accounts for such things as proximity effect of boundary surfaces or other cracks, orientation of the crack and the shape of the crack. For the simple case of a crack of length $2a$ in a large sheet subjected to a uniform stress $\delta_\infty$ remote from and perpendicular to the crack, $Y = 1$, and the stress intensity factor is given by:

$$K_\mathrm{I} = \sigma_\infty \sqrt{(\pi a)} \qquad (3)$$

The stress intensity factor $K_N$ can be related formally to the stress concentration factor $K_t$ by considering a crack as an elliptical notch of vanishingly small tip radius (see section 5.2.4.).

The usefulness of stress intensity factors in the analysis of problems of residual static strength, fatigue crack growth and stress corrosion is now well established; such problems form the substance of what is now called fracture mechanics. The power of the stress intensity factor method of analysis lies in the assumption that the behaviour of a sharp crack is determined by the stress field at the tip; it is thus necessary to determine the stress intensity factor only. This leads to a great simplification in the stress analysis necessary when the cracked structure is complex. After the foundations laid by Griffith (5/1, 5/2), Irwin (5/3) and Orowan (5/4) on fracture mechanics considerable effort has been devoted to evaluating $Y$ for a variety of problems (5/5, 5/6, 5/7, 5/8). The methods that have been used for evaluating $Y$ are briefly described in section 5.2 (theoretical methods)

---

* The numerical value of $Y$ will differ according to the definition used either $K = Y\sigma Wa$ or $K = Y\sigma W(\pi a)$ as here. See footnote to General notation on page vii. Editor.

and in section 5.3 (experimental methods). More detailed descriptions of these methods appear in the original papers and in reviews by Rice (**5/9**, Vol. 2), Sih (**5/10**) and Cartwright and Rooke (**5/11**).

## 5.2 THEORETICAL METHODS

### 5.2.1 Analytical

The methods considered here are those which satisfy all the boundary conditions exactly. Such methods have the advantage of leading to explicit expressions for stress intensity factors; but only certain classes of problems can be solved. In deriving the stress intensity factor use is made of the formal definition

$$K_N = \lim_{r \to 0} \sigma_N \sqrt{(2\pi r)} \tag{4}$$

where $\sigma_N$ is appropriate to the mode of cracking. For simplicity, all the methods are described for a crack of length $2a$ along the $x$ axis with the origin of the $(x, y)$ coordinates at the crack centre.

5.2.1(*a*) *Westergaard stress functions.* Westergaard (**5/12**) formulated in Airy stress function $F$, which for mode I, with self-equilibriated forces on the crack, takes the form

$$F_I = \text{Re}\,[\bar{\bar{Z}}_I] + y\,\text{Im}\,[\bar{Z}_I] \tag{5}$$

where $Z_I$ is the Westergaard stress function. $\bar{Z}_I$ and $\bar{\bar{Z}}_I$ are defined by

$$\frac{d\bar{\bar{Z}}_I}{dz} = \bar{Z}_I \text{ and } \frac{d\bar{Z}_I}{dz} = Z_I \tag{6}$$

where $z = x + iy$. The cartesian components of stress, in terms of $F_I$, are

$$\sigma_x = \frac{\partial^2 F_I}{\partial y^2},\ \sigma_y = \frac{\partial^2 F_I}{\partial x^2} \text{ and } \tau_{xy} = -\frac{\partial^2 F_I}{\partial x \partial y} \tag{7}$$

The simplest crack configuration studied by Westergaard was that of a crack in an infinite sheet subjected to uniform biaxial tension $\sigma_\infty$ at infinity; the stress function is

$$Z_I = \frac{\sigma_\infty z}{\sqrt{(z^2 - a^2)}} \tag{8}$$

Westergaard also studied a crack opened by wedge forces and an infinite series of collinear cracks under various loading conditions. The method can be extended to modes II and III and comparison of the stress field in terms of the Westergaard stress function $Z_N$ with equation (4) shows that the stress intensity factor is given by:

$$K_N = \sqrt{(2\pi)} \lim_{z \to a} \{\sqrt{(z-a)} Z_N\}, \ N = \text{I, II, III} \tag{9}$$

Several workers (**5/13, 5/14, 5/15**) have used Westergaard's method for solving crack problems.

**5.2.1(b) Complex stress functions.** Mushkelishvili's complex stress function approach (**5/16**) enables the Airy stress function $F$ to be written in terms of two complex functions $\phi(z)$ and $\psi(z)$, as

$$F = Re\ [z\phi(z) + \int \psi(z) dz] \tag{10}$$

which yields, from equation (7) with $F_I$ replaced by $F$

$$\sigma_x + \sigma_y = 4\ Re\ [\phi'(z)] \tag{11}$$

and

$$\sigma_x - \sigma_y + 2i\tau_{xy} = 2[z\phi''(z) + \psi'(z)] \tag{12}$$

From equation (11) and the known properties of $\phi'(z)$ (**5/16**) it can be shown that

$$K_I - iK_{II} = 2\sqrt{(2\pi)} \lim_{z \to z_1} \{\sqrt{(z-z_1)}\ \phi'(z)\} \tag{13}$$

This method of determining stress intensity factors has the advantage over the Westergaard method since conformal mapping can be used to map cracks into holes. For a mapping function

$$z = \omega(\zeta) \tag{14}$$

and a crack tip at $\zeta_1$ in the $\zeta$-plane, equation (13) becomes

$$K_I - iK_{II} = 2\sqrt{(2\pi)} \lim_{\zeta \to \zeta_1} \sqrt{\{\omega(\zeta) - \omega(\zeta_1)\}}\ \frac{\phi'(\zeta)}{\omega'(\zeta)} \tag{15}$$

where $\phi(\zeta) \equiv \phi(\omega(\zeta))$ and $\phi'(\zeta) = d\phi(\zeta)/d\zeta$. This method has been discussed in detail by Sih (**5/17**). It has been used, for example, by Erdogan (**5/18**) to obtain stress intensity factors for a crack in an infinite sheet in which point forces and moments act. The work of many Russian authors in this field has been collected by Savin (**5/19**). Many other examples are referenced by Paris and Sih (**5/5**).

Mode III deformation has been treated by Sih (**5/20**) and Shiryaev (**5/21**) using this method ($\phi'$ is replaced by the torsion function in (15)). A comparison of equations (9) and (13) shows that the Westergaard stress functions and the complex stress function are related by $Z_\text{I} - iZ_\text{II} = 2\phi'$.

### 5.2.2 Boundary collocation

In the method of boundary collocation the stress function (see section 5.2.1) is represented in series form and the determination of the stress intensity factor is reduced to the solution of a set of linear algebraic equations. The boundary conditions on the crack surface are built into the series chosen to represent the stress function, and any remaining conditions around finite boundaries are then fitted approximately. The boundary points may be matched (collocated) 'exactly' or fitted in a least squares sense. Whilst convergence is not guaranteed the boundary collocation method has contributed a considerable number of stress intensity factor solutions; accuracy of these solutions is assessed where possible by comparison with alternative solutions determined by analytical methods. Application of boundary collocation to crack problems starts either from the Williams' stress function or from series representation of complex stress functions.

*5.2.2(a) Williams' stress function.* The Williams' stress function (**5/22**) arises as a special case of the stress function for an infinite sector (**5/23**). It is an Airy stress function which also satisfies the conditions that the normal and shearing stresses be zero along the crack surface. It is convenient to write Williams' stress function as:

$$\chi = \chi_e + \chi_o \tag{16}$$

where $X_e$ and $X_o$ are the even and odd terms respectively and are given by

$$\begin{aligned}\chi_e = \sum_{n=1}^{\infty} \Bigg\{ &(-1)^{n-1} A_{2n-1}\, r^{n+\frac{1}{2}} \left[ -\cos\left(n - \frac{3}{2}\right)\theta \right.\\ &\left. + \frac{2n-3}{2n+1} \cos\left(n + \frac{1}{2}\right)\theta \right] + (-1)^n A_{2n} r^{n+1}\\ &\times [-\cos(n-1)\theta + \cos(n+1)\theta] \Bigg\}\end{aligned} \tag{17}$$

$$X_o = \sum_{n=1}^{\infty} \left\{ (-1)^{n-1} B_{2n-1} r^{n+\frac{1}{2}} \left[ \sin\left(n - \frac{3}{2}\right)\theta \right.\right.$$
$$\left. - \sin\left(n + \frac{1}{2}\right)\theta \right] + (-1)^n B_{2n} r^{n+1} \left[ -\sin(n-1) \right.$$
$$\left.\left. + \frac{n-1}{n+1} \sin(n+1)\theta \right] \right\} \tag{18}$$

$X_e$ and $X_o$ differ from equations (8) and (9) in reference (**5/22**) in which there is a typographical error. $r$ and $\theta$ are the usual crack tip polar coordinates. With the cylindrical polar form of equation (7) and the stress transformation equations from rectangular to polar coordinates, the stress intensity factor becomes, from equation (4),

$$K_{\mathrm{I}} = -A_1 \sqrt{2\pi} \tag{19}$$

and

$$K_{\mathrm{II}} = B_1 \sqrt{(2\pi)} \tag{20}$$

Boundary collocation of the Williams' stress function consists of the following stages:

(i) Determination of the Airy stress function $F$ for the uncracked configuration. These are often well known (**5/23**).

(ii) Evaluation of $F$ and the normal derivative $\partial F/\partial n$ at a number of points (collocation points) around the boundary.

(iii) Substitution of $F$ and $\partial F/\partial n$ at each point into the Williams' stress function $\chi$ and its derivative $\partial \chi/\partial n$ to yield a set of linear simultaneous equations the solution of which gives $A_1$ and $B_1$.

The test of convergence used is to increase the number of matched points (typically 20 to 30) until $A_1$ and $B_1$ reach a stable value. Considerable use has been made of boundary collocation of the Williams' stress function by Gross et al (**5/24 – 5/28**) for conditions of geometry and loading which are symmetrical with respect to the plane of the crack. Rather than fitting $F$ and $\partial F/\partial n$ exactly it can be done in a least squares manner. Boundary collocation of the Williams' stress function is suitable for singly connected configurations in two-dimensional problems.

The Williams' stress function (5/29) as modified by Sih and Rice (5/30) has been used with boundary collocation to obtain stress intensity factors for finite bi-material plates (5/31). The method was used to analyse a centrally cracked bi-material plate and a partially debonded composite laminate.

5.2.2(b) *Complex stress functions.* Stress intensity factors in multiply connected two-dimensional bodies subjected to in-plane loading may be determined from a series representation of the complex functions $\phi(z)$ and $\Omega(z)$ (see Muskhelishvili (5/16)). The stresses and displacements in terms of these functions (see also section 5.2.2.(a)) are

$$\sigma_x + \sigma_y = 2[\phi'(z) + \overline{\phi'(z)}] \tag{21}$$

and

$$\sigma_y - \sigma_x + 2i\tau_{xy} = 2[(\bar{z} - z)\phi''(z) - \phi'(z) - \overline{\Omega(z)}] \tag{22}$$

To ensure single valued displacements around the crack we require:

$$\kappa \oint_C \phi'(z)dz - \oint_C \Omega(\bar{z})d\bar{z} = 0 \tag{23}$$

where $C$ is taken around each internal boundary and $\kappa$ is a function of Poisson's ratio. It is assumed that $\phi'$ and $\Omega$ can be expanded in series form, i.e.

$$\phi'(z) = \frac{1}{\sqrt{(z^2 - a^2)}} \sum_{n=0}^{\infty} C_n z^n + \sum_{n=0}^{\infty} D_n z^n \tag{24}$$

and

$$\Omega(z) = \frac{1}{\sqrt{(z^2 - a^2)}} \sum_{n=0}^{\infty} C_n z^n - \sum_{n=0}^{\infty} D_n z^n, \tag{25}$$

which automatically satisfy the boundary conditions on the crack. The stress intensity factor is given by equation (13).

The method of solution is to determine a finite number of coefficients $C_n$ and $D_n$ by collocating the prescribed stresses around the boundary, taking into account the condition for single valued displacements. These procedures have been used by Kobayashi *et al* (5/32), Vooren (5/33) and Newman (5/34). The probable errors in their results were estimated from the effects of including more terms in the series for complex stress functions.

Isida (**5/35**–**5/40**) has made extensive use of series representation of complex functions. Results of many problems involving mode I, II and III deformation, interaction between cracks and cracks in stiffened sheets are reported in reference (**5/10**). In the method used by Isida series are deveoped for a stress-free elliptical boundary and the stress intensity factor is determined using either equation (13) or (26) in the limiting case as the ellipse becomes a crack.

### 5.2.3 Conformal mapping

Conformal mapping has been mentioned in section 5.2.1.(b). For complicated boundaries, e.g. radial cracks emanating from a circular hole (**5/41**), the exact form of the mapping function (equation 14) becomes multivated and subsequent stress analysis is difficult. Bowie (**5/41**–**5/43**) has developed a technique for avoiding these difficulties by representing the mapping function as a polynomial, the coefficients of which are determined by comparison with the exact mapping function. This method is applicable to configurations for which an exact mapping is available; it has been applied by several workers to singly connected regions (**5/44**–**5/47**). A method in which conformal mapping is combined with boundary collocation (section 5.2.2.(b)) has been developed by Bowie and Neal (**5/48**). This method avoids the difficulty of finding accurate polynomial mapping functions of the complete physical region. A simple form of mapping function is used to transform a crack and its exterior in the physical plane into a circle and its exterior in the auxiliary plane. The remainder of the boundary in the physical plane corresponded to a directly calculable curve in the auxiliary plane. Collocation methods are used to satisfy conditions around this boundary. Bowie and Freese (**5/49, 5/50**) have applied the modified mapping-collocation technique to doubly connected regions in both isotropic and orthotropic sheets. Recently the method has been extended (**5/51**) to improve convergence in certain complicated configurations.

### 5.2.4 Stress concentrations

Irwin (**5/14**) proposed that Neuber's (**5/52**) stress concentration factors for notches of small flank angle and small root radius may be used to obtain theoretical expressions for stress intensity factors. Consider a notch which in the limit of zero root radius ($\rho$) tends to a crack along the $x$-axis: if $\sigma_{max}$ is the maximum value of $\sigma_y$ at the tip we have:

$$K_I = \frac{\sqrt{\pi}}{2} \lim_{\rho \to 0} \{\sigma_{max} \sqrt{\rho}\} \tag{26}$$

if $\tau_{max}$ is the maximum value of $\tau_{xy}$ near the tip along the $x$-axis (the position of the maximum tends to the tip as $\rho \to o$) we have

$$K_{II} = \sqrt{\pi} \lim_{\rho \to 0} \{\tau_{max} \sqrt{\rho}\} \tag{27}$$

and similarly if $\tau_{max}$ is the maximum value of $\tau_{zy}$ at the tip we have

$$K_{III} = \sqrt{\pi} \lim_{\rho \to 0} \{\tau_{max} \sqrt{\rho}\} \tag{28}$$

Although the relationship between $K_N$ and $\sigma_{max}$ (or $\tau_{max}$) is exact the actual expressions for the maximum stresses may be known only approximately. Harris (5/53) has made considerable use of equations (26) to (28) and Neuber's work on stress concentration factors in deriving expressions for $K_I$, $K_{II}$ and $K_{III}$ in circumferentially cracked round bars under bending, transverse shear, torsion and longitudinal tension. Pook and Dixon (5/54) have analysed a finite rectangular sheet with an edge crack loaded such that there is combined tension and bending at the crack tip.

### 5.2.5 Green's functions

Certain solutions (5/55, 5/18) are particularly useful in that they may be used to construct Green's functions for determining stress intensity factors for cracks in arbitrary stress fields (5/5). The stress intensity factor, for the tip at $x = a$, due to a pair of point forces $Q$ perpendicular to the crack surface ($-a < x < a$, $y = 0$) and at a distance $x$ from the origin, is given by

$$K_I = Q \left[ \frac{1}{\pi a} \left( \frac{a+x}{a-x} \right) \right]^{\frac{1}{2}} \tag{29}$$

the coefficient of $Q$ is the Green's function.

If $\sigma_y(x, o)$ is the stress along the crack site in the absence of the crack, then writing $Q = \sigma_y(x, o)dx$ and integrating over the crack length gives the contribution of all the forces distributed along the crack line, i.e.

$$K_I = \frac{1}{\sqrt{(\pi a)}} \int_{-a}^{+a} \left[ \frac{a+x}{a-x} \right] \sigma_y(x, o)dx \tag{30}$$

$\sigma_y(x, o)$ may be measured experimentally in the uncracked solid or determined theoretically. Equation (30) then enables the stress intensity factor to be determined. Equivalent forms exist for modes II and III (**5/5, 5/56**).

For a penny-shaped crack the axisymmetric Green's function $g(r)$ is given (**5/57**) by

$$g(r) = \frac{2}{\sqrt{(\pi a)}} \cdot \frac{r}{\sqrt{(a^2 - r^2)}} \tag{31}$$

where $r$ is the distance from the centre of the crack, of radius $a$. Thus the stress intensity factor due to a pressure $p(r)$ is given by

$$K_I = \frac{2}{\sqrt{(\pi a)}} \int_0^a \frac{rp(r)}{\sqrt{(a^2 - r^2)}} \, dr \tag{32}$$

Graphical and analytical Green's functions have been determined by several workers (**5/8, 5/10, 5/58**) and applications to several problems of practical interest can be found in the work of Chell (**5/59**).

### 5.2.6 Integral transforms and dislocation models

In this method the elastic problem is considered as a mixed boundary value problem and solved using standard transform techniques. Many solutions using various transforms have been collected by Sneddon and Lowengrub (**5/60**). These methods usually reduce to the solution of an integral equation which takes the form

$$\int_{-a}^{a} K(s, x)q(s)ds = L(x) \tag{33}$$

where $L(x)$ is a function of the known stress along the crack site in the uncracked body, $K(s, x)$ is the known kernel, and $q(s)$ is the unknown function. The function $q(x)$ is proportional to the derivative of the relative displacement $v(x, o)$ of the crack faces, i.e.

$$\frac{E}{2(1 - v^2)} \frac{\partial v(x, o)}{\partial x} = q(x) \tag{34}$$

where $E$ is Young's modulus and $v$ is Poisson's ratio.

Thus a knowledge of $q$ enables the stress intensity factor to be determined from the relationship

$$K_a = \lim_{x \to a} \left[ \sqrt{(2\pi(a - x))} \frac{E}{2(1 - v^2)} \frac{\partial v(x, o)}{\partial x} \right] \tag{35}$$

Analytic solutions have been found for several two-dimensional and three-dimensional crack problems in the infinite domain (**5/60**). With complex crack shapes and boundaries the integral equations (33) may require numerical solutions, e.g. Gaussian quadrature. Rooke and Sneddon (**5/61**) used a series expansion of $q$ to solve the problem of a star-shaped crack subjected to an internal pressure. A development by Tweed (**5/62**) enables certain plane problems, e.g. radial cracks in discs and cylinders to be solved using Mellin transforms (**5/63–5/65**). Smetanin (**5/66**) has solved the problem of an annular crack in a uniaxial tensile field and problems of cracks in infinite strips have been considered by Fichter (**5/67**).

A method based on continuous dislocation arrays also leads to integral equations similar to equations (33).

Cracks are represented as a distribution of dislocations. The density of the distribution is determined by satisfying boundary conditions. For cracks within elastic continua a Burger's vector of $D_i(s)ds$ is used as the Burger's vector of an infinitesimal dislocation of density $D_i(s)$ at location $s$. By analogy with the definition of a physical dislocation, the Burger's vector of a continuous array of dislocations is given by:

$$b_i = \int_s D_i(s)ds \tag{36}$$

where $S$ is a path around the dislocation distribution.

Consider a crack in an infinite sheet opened by an internal pressure $p(x)$. The opening of the crack can be represented by a continuous array of dislocations of density $D_y(s)$ lying along $y = 0$, $|x| < a$ and having a Burger's vector $b_Y = \int_s D_y(s)ds$. These dislocations cause stresses along $y = 0$ of

$$\sigma_y = -\sigma_x = \frac{2\mu}{\pi(1+\kappa)} \int_{-a}^{+a} \frac{D_y(s)ds}{(x-s)}, \tau_{xy} = 0 \tag{37}$$

where $\mu$ is the shear modulus and $\kappa = (3 - 4v)$ in plane strain or $(3 - v)/(1 + v)$ in plane stress. The boundary conditions on $y = 0$ are $\sigma_y = -p(x)$ for $|x| < a$ and $\tau_{xy} = 0$ for all $x$. These conditions can be satisfied, using equation (37) by ensuring that

$$p(x) + \frac{2\mu}{\pi(1+\kappa)} \int_{-a}^{+a} \frac{D_y(2)ds}{(x-s)} = 0 \tag{38}$$

For the displacements to be single valued around a circuit enclosing the crack there must be no net Burger's vector around the crack, i.e.

$$\int_{-a}^{+a} D_y(s) \, ds = 0 \tag{39}$$

The solution of equations (38) and (39) for $D_y(s)$ enables the stress intensity factor to be determined from the relation

$$K_{\mathrm{I}} = \frac{2\sqrt{2\pi}\mu}{(1+\kappa)} \lim_{x \to a} \{(a-x)^{\frac{1}{2}} D_y(x)\} \tag{40}$$

Equivalent formulations can be made for modes II and III. For more complex configurations equation (37) will contain terms which account for the interaction of the dislocation with additional boundaries, for example see Atkinson (5/68, 5/69) and Rice (5/9, Vol. 2). Bilby and Eshelby (5/9, Vol. 1) have reviewed the application of dislocation models to the representation of cracks.

A versatile method, similar to the dislocation approach has recently been proposed (5/70) in which the stress field due to a point force singularity is used to form a crack. In this method distributed point forces are applied along the segment at which a crack is to be formed. The unknown density of these forces is determined such that the boundary conditions are satisfied on the segment of the crack. The method can be used for both two-dimensional and three-dimensional problems providing that the Green's function for a point force in the uncracked configuration is known.

### 5.2.7 Force-displacement matching

This method is applicable to configurations in which there are boundaries between different materials; it involves matching the forces and displacements in the region over which the materials are joined. The method has been used mainly for determining stress intensity factors for cracks in stiffened sheets (**5/71–5/80**) where the stiffener is represented as a distribution of point forces. The distribution is determined by satisfying compatibility of displacements and equilibrium of forces at each attachment point between the sheet and the stiffener. In the case of discretely attached stiffeners (**5/71–5/78**) the forces occur at a series of points (e.g. rivet locations). These forces are obtained by solving a series of about thirty simultaneous equations. With continuously attached stiffeners (**5/79, 5/80**) the forces would be distributed along the stiffener/sheet junction and this distribution is determined by reducing the resulting integral equation (see section 5.2.6) to a series of simultaneous equations and solving as for discretely attached stiffeners. In both the continuously attached and discretely attached stiffener configurations, the stress intensity factor is determined by summing the effects due to the forces by using a known Green's function for a single force (see section 5.2.5). In formulating the compatibility condition it is possible to allow for relative displacements between the sheet and the stiffener at the attachment points such as would occur with shear deflection of rivets or adhesives (**5/79**).

### 5.2.8 Alternating methods

This method, sometimes referred to as the Schwarz alternating technique (**5/81**) has been useful in determining stress intensity factors for a number of two-dimensional and three-dimensional crack problems and many applications have been reviewed by Hartranft and Sih (**5/10**, Ch. 4). The alternating method involves knowing the solution of, usually, two auxiliary problems and is most useful in determining the effect of a single stress-free boundary near to or intersecting the crack. When applied to crack problems one auxiliary solution (first solution) will be for a loaded crack in an infinite plane and the other (second solution) will be for a plane containing a boundary ($B$) subjected to an arbitrary stress distribution. From the known boundary conditions on the crack, the first solution is used to calculate stresses along a line ($L$) where the stress-free boundary is to be created. The second solution is then superimposed with the stresses on $B$ adjusted to cancel those existing along line $L$. This makes $L$ stress-free but introduces residual stresses on the crack surfaces. The first solution is used to cancel the residual stresses which causes new, but reduced stresses on the line $L$. The second solution is again used and the alternating procedure is repeated until the known boundary conditions on the crack are satisfied and the stresses on $L$ become negligible. The stress intensity factor is obtained by summing the effect of removing the stresses from the crack, at each stage of the alternating procedure, using the known Green's function (see section 5.2.5). The method has been used extensively for three-dimensional problems by Smith *et al* (**5/82–5/86**) for part-circular cracks, by Shah and Kobayashi (**5/87–5/89**) for elliptical cracks.

### 5.2.9 Finite elements

The methods for determining stress intensity factors using finite elements (**5/90**) fall into three categories. Firstly those which allow stress intensity factors to be determined directly, secondly those which require stress intensity factors to be determined indirectly by considering changes in energy due to the presence of the crack, and thirdly those involving special crack tip elements. The direct methods require small elements in the vicinity of the crack tip. The indirect methods do not require such small elements in the vicinity of the crack tip but suffer a loss of accuracy resulting from differentiation. Some of the indirect methods, e.g. compliance and strain energy methods, have the disadvantage that $K_I$ and $K_{II}$ in mixed mode situations cannot be separated.

In an attempt to avoid these disadvantages a number of methods involving special elements in the vicinity of the crack tip have been developed. The various methods which have been used for determining stress intensity factors by finite elements are treated separately in the following sections: more detailed discussions of the methods can be found in reviews by Wilson (5/10, Ch. 9). Jerram and Hellen (5/91) and Oglsby and Lomacky (5/92).

5.2.9(a) *Crack tip stress and displacement.* The stress method correlates the stresses at the nodal points of the finite element mesh with those at the crack tip which are given by:

$$\sigma_{ij}(r, \theta) = \frac{K_N}{\sqrt{(2\pi r)}} f_{ij}(\theta) \tag{41}$$

where $r$, $\theta$ are polar co-ordinates centred at the crack tip and $f_{ij}(\theta)$ is a known function of $\theta$ (5/5). For the opening mode ($N = I$) equation (41) can be written:

$$K_I = \sigma_y(r, o) \sqrt{(2\pi r)} \tag{42}$$

Hence $K_I$ may be determined from $\sigma_y(r, o)$ at some small distance $r$ from the crack tip. The method has been used by Chan et al (5/93) and Kobayashi et al (5/94). Both authors conclude that the results are not as accurate as those obtained by correlating the displacements of the finite element nodal points. For the opening mode ($N = I$)

$$K_I = \frac{Eu_y}{4(1 - v^2)} \sqrt{\frac{2\pi}{r}} \tag{43}$$

hence $K_I$ may be determined from the crack surface displacement $u_y$ at some small distance $r$ from the crack tip. The method has been used by many workers (5/94–5/97) for a variety of problems.

5.2.9(*b*) *Energy methods.* These methods do not need such small elements in the vicinity of the crack tip. In the method proposed by Hayes (**5/98**), the energy integral of Bueckner (**5/99**) is used to calculate opening mode stress intensity factors.

$$W = \frac{1}{2}\int_{S} \sigma_{ij}u_{i}\mathrm{d}s \qquad (44)$$

in which $\sigma_{ij}$ is the distribution of stress along the line on which the crack is to be introduced, $u_i$ is the displacement of the crack face due to $\sigma_{ij}$ acting on the crack face (the remainder of the boundaries being stress free), and $S$ is an integration path over the entire crack surface. The stress intensity factor is determined by evaluation $W$ at a number of crack lengths, numercially differentiating $W$ with respect to crack length to get the strain energy release rate $G$ and hence $K$ from the known relationship (**5/5**). The stress $\sigma_{ij}$ may often be determined from known analytical solutions. In more complicated cases numerical methods may be required. If it is necessary to determine $\sigma_{ij}$ from a finite element analysis then elements need only be small relative to geometric stress concentrations and small elements are not needed for the determination of $u_i$. A similar method has been used by Jerram (**5/100**) in which he calculated the work to close a small segment of crack. Recent work of Hellen (**5/101**) avoids the need to difference the energy between two separate calculations by using a virtual crack extension technique. These energy methods are applicable to proalems involving thermal stresses, body forces, residual stresses and mixed mode deformation. Other, less versatile, energy methods have been proposed. These involve differencing the total energy of the structure over successive crack lengths (**5/95, 5/101, 5/102**), determining the rate of change of the compliance of the structure with crack length (**5/103–5/106**) (see section 3.1), and evaluating a line integral (**5/9**, Vol. 2) around a contour surrounding the crack tip (**5/93**).

5.2.9(*c*) *Special crack tip elements.* Many types of special element have been developed for obtaining stress intensity factors and applied to a variety of configurations (**5/107–5/115**).

These special element methods enable stress intensity factors to be determined directly and, in general, they require less elements than the methods previously described. The special elements usually require that the analytical form of the singularity be known, and they also involve some modification of standard finite element programmes.

## 5.3 EXPERIMENTAL METHODS

In this section some experimental methods of determining stress intensity factors are described. Experimental methods may either use a known relationship between a measurable quantity (e.g. compliance or fatigue crack growth rate) and the stress intensity factor, or involve direct measurements on a model (e.g. by photoelasticity).

### 5.3.1 Compliance

Irwin and Kies (5/116) showed that the strain energy release rate $G$ could be written in terms of the applied load $Q$ and the change in compliance $C$ with respect to crack area $A$ as

$$G = \frac{Q^2}{2} \frac{dC}{dA} \qquad (45)$$

from which the stress intensity factor for plane stress is given by

$$K_I = Q \left[ \frac{E}{2} \frac{dC}{dA} \right]^{\frac{1}{2}} \qquad (46)$$

The determination of $K_I$ from equation (46) involves measuring the compliance $C$ for a range of crack lengths. The derivative of the compliance vs. crack length curve enables $K_I$ to be found. The experimental method needs considerable care if satisfactory results are to be obtained (see the work of Srawley et al (5/117)). The method has been used for many problems (5/117–5/121), more details of the technique can be found in reviews by Bubsey et al (5/112. Ch. 4) and Cartwright and Rooke (5/11).

### 5.3.2 Photoelasticity

Of the optical methods for determining stress intensity factors photoelasticity has been most used. The technique has several advantages. It is a well-known method for which experimental equipment and birefringent materials are readily available. By using the frozen stress technique photoelastic analysis may be extended to three dimensional configurations; an assessment of this technique in relation to crack tip stress fields has been made by Schroedl *et al* (**5/123**). Considerable use has been made of photoelasticity in determining stress concentration factors although the technique has not yet been used to the same extent to determine stress intensity factors: reviews have been carried out by Kobayashi (**5/122**), Hardy (**5/124**) and Marloff *et al* (**5/125**). Use of photoelasticity for crack problems inevitably involves representing the crack by a narrow slit of finite root radius. This creates a need to define an equivalent crack length (**5/91, 5/126**). The various photoelastic methods which have been used to determine stress intensity factors will now be described.

Two methods, which involve measurements of the stress near to a simulated crack, are based on the equation for the maximum shear stress $\tau_m$, given by

$$\tau_m = \frac{1}{2\sqrt{(2\pi r)}} [(K_I \sin\theta + 2K_{II} \cos\theta)^2 + (K_{II} \sin\theta)^2]^{\frac{1}{2}} \tag{47}$$

where $(r, \theta)$ are polar coordinates centred at the crack tip. In the method used by Emery *et al* (**5/127**) the stress intensity factors were evaluated from the maximum shear stress $\tau_m$ on lines perpendicular to and through the crack tip given by:

$$\tau_m = \frac{(K_I^2 + K_{II}^2)^{\frac{1}{2}}}{2\sqrt{(2\pi r)}} \tag{48}$$

and on a line outside and collinear with the crack given by

$$\tau_m = \frac{K_{II}}{\sqrt{(2\pi r)}} \tag{49}$$

Evaluation of $\tau_m$ close to the tip at points above (or below) and ahead of the notch allows both $K_I$ and $K_{II}$ to be determined from equations (48) and (49). In the method used by Smith and Smith (**5/128, 5/129**) equation (47) is used together with the condition $\partial \tau_m/\partial\theta = 0$, that is the condition

$$\left(\frac{K_{II}}{K_I}\right)^2 - \frac{4}{3}\left(\frac{K_{II}}{K_I}\right) \cot 2\theta_m - \frac{1}{3} = 0 \tag{50}$$

The angle $\theta_m$, is that at which a tangent to the isochromatic fringes is perpendicular to the radius $r$. This angle is measured from the isochromatic fringe pattern near the tip, the ratio $K_{II}/K_I$ is determined from equation (50) and hence $K_I$ and $K_{II}$ can be found using a known value of $\tau_m$. In equation (47). Both these methods have the advantage that $\tau_m$ may be obtained directly from the isochromatic fringe pattern, and so avoid the need for stress separation.

Wilson (**5/130**) used the crack-tip stress field, along the crack line

$$\sigma_y = \frac{K_\mathrm{I}}{\sqrt{(2\pi)}} \cdot x^{-\frac{1}{2}} \tag{51}$$

to obtain the stress intensity factor for a WOL specimen from the photoelastic results of Leven (**5/131**); a plot of $\sigma_y$ against $x^{-\frac{1}{2}}$ should be a straight line of slope $K_\mathrm{I}/\sqrt{(2\pi)}$. The stress intensity factor was also calculated from the average stress over a small distance from the crack tip:

i.e. 
$$K_\mathrm{I} = \sqrt{\left(\frac{\pi}{2x_1}\right)} \int_0^{x_1} \sigma_y \, dx \tag{52}$$

This procedure has the advantage of reducing the random scatter in experimental data and results in a more accurate value of the stress intensity factor. In both the methods used by Wilson (**5/130**) a stress separation technique must be used to find $\sigma_y$.

Another method, which makes use of an approximate form of the relationship between the maximum stress at the tip of a narrow notch and the stress intensity factor when the notch becomes a crack (see section 5.2.4),

$$K_\mathrm{I} \simeq \frac{\sqrt{(\pi\rho)}}{2} \sigma_{\max} \tag{53}$$

i.e. has been used by Wilson (**5/130**) and Pook and Dixon (**5/54**). Since the minimum principal stress is zero at the tip $\sigma_{\max}$ can be determined directly from the photoelastic measurements. However, since the stress gradient is steep in the tip region, accurate determinations of $\sigma_{\max}$ are difficult; it is also necessary to measure the radius of the notch tip $\rho$ accurately.

### 5.3.3 Fatique crack growth rate

Paris (**5/132**) proposed that the growth rate of a crack extending under fatigue loading could be characterized by the stress intensity range

$$\frac{da}{dN} = f(\Delta K) \tag{54}$$

where $da/dN$ is the crack growth rate and $\Delta K$ the stress intensity factor range, i.e.

$$\Delta K = K_{\max} - K_{\min} \tag{55}$$

To determine the stress intensity factor for a new configuration it is necessary to conduct a fatigue test on that configuration and record both the length and the rate of growth of the crack over the range of crack lengths required. A fatigue test must then be performed under identical conditions on a specimen of the same material, for which the stress intensity factor is known as a function of stress and crack length. The data from the two tests are then compared on the basis of equivalent crack growth rates. James *et al* (**5/133**) have applied this method to the case of a thick walled, internally pressurized, cylinder containing a longitudinal crack. This method has the advantage that stress intensity factors may be determined under conditions very close to those existing in the real structure.

### 5.3.4 Interferometry and Holography

Two optical methods based on measurements of interference fringes in transparent materials have been proposed. The method of Dudderar and Gorman (**5/134**) involves determining the opening mode stress intensity factor from measurements of the magnitude of the sum of the normal stresses at a series of points outside and collinear with the crack. A thin perspex sheet, containing a sharp notch is subjected to a tensile stress. The magnitude of $(\sigma_x + \sigma_y)$ is obtained, for several loads from a series of interferograms. The magnitude of $K_\text{I}$ is calculated from the slope of a plot of $(\sigma_x + \sigma_y)$ vs. $r^{-\frac{1}{2}}$ in a similar way to that used by Wilson (**5/130**) for photoelastic results (see Section 3.2). The method, proposed by Sommer (**5/135**) involves measuring the relative displacement of the crack faces, by an interference technique, in glass sheets under load. The relative displacement of the crack faces is determined at several positions along the crack and the opening mode stress intensity factor obtained using the known relationship between the displacements and $K_\text{I}$.

## 5.4 CONCLUDING REMARKS

The methods described have been developed during the last thirty years and have contributed over 250 solutions of which approximately 80 per cent have been for two-dimensional configurations. Providing that a computer is available the numerical methods for evaluating stress intensity factors are sufficiently well developed to enable most problems to be solved in two-dimensions. For problems in three-dimensions the methods available are more limited; techniques which involve using distributions of point forces, e.g. the alternating method, are restricted to configurations for which basic solutions are available. The finite element method shows the greatest potential for solving three-dimensional problems. This method has the highest growth rate and, with the introduction of special crack tip elements, it is likely that it will be used more extensively in the future. Of the experimental methods, only the optical (sections 5.3.2 and 5.3.4) can be used for analysing both mixed mode and three dimensional configurations; the maximum shear stress method is the best of the photoelastic methods presently available. The fatigue crack growth rate method has the advantage that tests can be conducted under service conditions.

The choice of a means of evaluating a stress intensity factor will often depend on the accuracy required. The methods based on boundary collocation, conformal mapping, integral transforms and force-displacement matching are all accurate to better than 2 per cent; methods which reduce to an integral equation can often be made more accurate than this. Alternating methods and those based on the finite element technique may contain errors up to 6 per cent. The experimental methods are, in general, less accurate. Errors can be kept within 2 per cent–5 per cent using the compliance method providing great care is exercized. The optical methods may contain errors of the order of 10 per cent.

# 6
# Experimental Methods for Fracture Toughness Measurement

A. H. PRIEST

*British Steel Corporation, Sheffield Laboratories, Swinden House, Moorgate, Rotherham*

Recommended methods for measuring fracture toughness parameters are discussed. The relevant British Standards are:
  (a) BS5447: 1977 'Methods of Test for Plane Strain Fracture Toughness ($K_{Ic}$) of Metallic Materials'.
  (b) Draft for Development 19: 1972 'Methods for Crack Opening Displacement (COD) Testing'.

The latter document is being revised and up-dated for publishing as a full British Standard and probable additions and alterations are noted.

The situation with respect to non-standard tests is reviewed. The fracture toughness parameters under this heading include $J_{Ic}$, equivalent energy values, fracture propagation energy values and $R$-curve analysis. Particular attention is paid to the influence of strain rate on testing procedures and a crack monitoring technique is discussed.

## 6.1 INTRODUCTION

The two testing procedures at present used in the United Kingdom for fracture mechanics tests are BS5447: 1977 'Methods of Test for Plane Strain Fracture Toughness ($K_{Ic}$) of Metallic Materials' and Draft for Development 19: 1972 'Methods for Crack Opening Displacement (COD) Testing'. The latter document is being revised and should be available as a full British Standard in the near future.

*The MS of this paper was received at the Institution of Mechanical Engineers on 9 June 1975 and accepted for publication on 22 August 1975.*

# METHODS FOR FRACTURE TOUGHNESS MEASUREMENT

The BSI procedures have been devized so as to permit measurement of force ($K_{Ic}$) and displacement (COD) under identical conditions with the same laboratory equipment and in the same test.

The unified testing procedure is:

1. A machined notched test piece is used in which a tensile stress is developed at the notch tip either by tension or by bending.

2. The test piece is subjected to repeated loading, the magnitude and number of load cycles being chosen to start a sharp crack and develop it to the desired length.

3. The test piece is then loaded in tension or bending and a record is taken of the load against the opening displacement of the two sides of the notch. The test is continued until fracture occurs.

4. The fracture toughness parameters are calculated from the load, displacement and initial crack length.

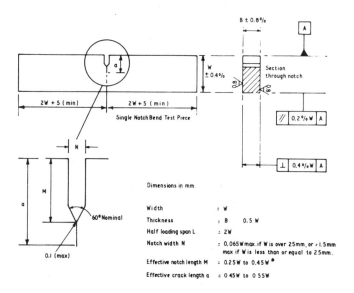

*Fig. 18 Proportional dimensions and tolerances for bend test pieces.*

## 6.2 STANDARD METHODS

### 6.2.1 $K_{Ic}$ testing

*6.2.1(a) Types of test piece.* Two main types of test piece are recommended by BSI:

(i) *Bend test pieces*

These are the simplest and the cheapest to machine. Limitatons are imposed upon the test piece dimensions and the shape and depth of the notch. The pieces most commonly used have $B = \frac{1}{2}W$, $a = 0.45W$ to $0.55W$ and $M = 0.25W$ to $0.45W$, Fig. 18.

(ii) *Compact tension test pieces*

Tension test pieces use less material, but the cost of machining the clevis pinholes accurately is considerable. The recommended dimensions are $B = \frac{1}{2}W$, $a = 0.45W$ to $0.55W$ and $2F = 0.275W$, Fig. 19.

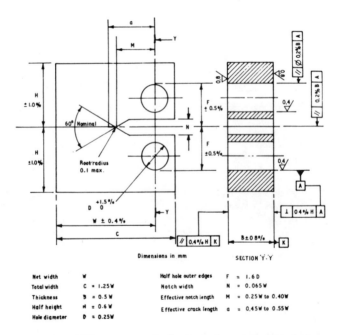

*Fig. 19 Proportional dimensions and tolerances for tensile test pieces.*

6.2.1(*b*) *Testing fixtures.* To reduce friction to a minimum, bend tests are made on freely rotating rollers with or without fixed centres. The diameter of the rolls and centre loading point should be between $\frac{1}{3}W$ and $W$. For tension tests a clevis and pin allows rotation and self-alignment as the test piece is loaded, and a fairly loose fit between clevis pin and clevis holes is advisable. Details of recommended designs for the loading fixtures are given in an Appendix to the proposed standard.

6.2.1(*c*) *Test piece dimensions.* The plane strain fracture toughness of a material is a measure of its resistance to unstable fracture under conditions where plastic deformation is limited to a region of the crack tip which is small in comparison to the test piece size, but the manner and the degree in which plasticity is to be limited has been the subject of much research and the influence of size of test piece in this regard is still under debate.

In their early work Srawley and Brown (**6/1**) limited plastic deformation simply by restricting the nominal stress at the crack tip to below the yield stress. The nominal stress is the stress that would exist at the crack tip if there were no stress intensification there. In a centre cracked tension test piece the nominal stress is the average stress in the unbroken ligament:

$$\sigma_{nom} = \sigma_{nett} = \frac{Q}{B(W-a)} \qquad (1)$$

In a compact tension test piece:

$$\sigma_{nom} = \frac{2Q\,(2W+a)}{B\,(W-a)^2} \qquad (2)$$

and in a bend test piece:

$$\sigma_{nom} = \frac{3QL}{B\,(W-a)^2} \qquad (3)$$

Srawley and Brown defined the measurement capacity of each type of test piece as the maximum value of fracture toughness that could be measured before the nominal stress to cause yielding was exceeded, and they determined these values for several values of $a/W$ for single edge notch tension test pieces having different loading axis positions, for centre cracked tension test pieces, and for three and four point bend test pieces.

For most types of test piece the highest values were found at $a/W$ values between 0·2 and 0·4 and for bend test pieces these values proved about 15 per cent higher than those of tension test pieces. In all these tests the measurement capacity was proportional to the test piece width, $W$.

These observations alone sufficed to establish good practice with $a/W$ values of between 0·2 and 0·4 and led the ASTM special committee on Fracture Testing (6/2) to adopt as one of their first recommendations that the nominal stress should not exceed 80 per cent of the 0·2 per cent proof stress.

Another requirement is to limit plastic deformation by maintenance of plane strain conditions at the crack tip. High triaxial stresses are developed when the test piece is thick enough to preclude yield at the surface spreading laterally to the centre of the crack front. In many materials such yielding is accompanied by the formation of shear lips on the fracture surface and if shear lips occupy the whole of the fracture surface then plane strain conditions are not present. However, absence of shear lips is not to be regarded as proof that plane strain conditions prevail.

To ensure plane strain it was earlier suggested (6/3) that the thickness, $B$, should exceed four times the plastic zone size, $r$, at the crack tip, i.e.

$$B > \frac{2}{\pi} \left(\frac{K}{\sigma_Y}\right)^2 \tag{4}$$

In order to afford a proper basis for standardization of test procedures for measurement of fracture toughness ASTM E24 and BISRA MG/EB Committees instituted collaborative test programmes on a variety of materials including steels, aluminium and titanium alloys; test pieces with a wide variety of dimensions including crack length, thickness and width were tested.

The results of these programmes indicated that consistent $K_{Ic}$ values could be obtained if both crack length, $a$, and the thickness of test pieces, $B$, were greater than 2·5 times the ratio $(K_{Ic}/\sigma_Y)^2$, moreover $a/W$ values greater than 0·55 were considered undesirable because at high $a/W$ values small errors in crack length can give rise to large errors in $K_{Ic}$.

6.2.1(*d*) *Fatigue pre-cracking*. The collaborative programme led to requirements for the fatigue pre-cracking operation. These are:

(a) The ratio of minimum to maximum force shall be in the range 0 to 0·1.

(b) $K_f$ shall not exceed 0·7 $(\sigma_{Y1}/\sigma_{Y2})\ K_Q$ where $\sigma_{Y1}$ and $\sigma_{Y2}$ are the proof stresses at the temperatures of fatigue cracking and at the test temperature respectively.

(c) The crack length shall be not less than 1·25 mm and the $a/W$ ratio shall be in the range 0·45 to 0·55.

(d) The crack shall lie within a defined envelope.

After testing the crack is measured and the effective crack length is calculated from the average of three measurements at 25, 50 and 75 per cent $B$. Crack curvature, the angle of the crack plane and the development of multiple cracks are also subject to limitations.

These requirements and limitations differ slightly from those recommended by ASTM but the differences are considered insignificant.

6.2.1(*e*) *Instrumentation.* A fracture toughness test is conducted in much the same way as a tensile test, the extensometer to measure strain being replaced by a guage located across the notch of the test piece to measure the crack opening displacement. The clip gauge recommended is a double cantilever with strain gauges on all four surfaces (Fig. 20). Force and displacement should be determined with an accuracy of better than $\pm 1$ per cent if the offset procedure discussed below is to be used.

Grooves at the extremities of the arms of the displacement gauges are seated on knife edges firmly attached to the test piece on either side of the notch. For tests at low or high temperatures a clip gauge extensometer (**6/4**) may be needed, Fig. 20.

Four typical force displacement diagrams are shown in Fig. 21.

6.2.1(*f*) *The offset procedure.* $K_{Ic}$ values were originally calculated from the pop-in load (**6/5**) illustrated in Fig. 21, Type IV. However because many materials do not exhibit pop-in it became necessary to establish a procedure to enable $K_{Ic}$ values to be determined from force-displacement records showing deviations from linearity but no sudden discontinuity, i.e. Type I, Fig. 21. The secant intercept procedure was adopted as a means of separating deviations from linearity due to plasticity from those due to crack growth.

This procedure is as follows: Referring to Fig. 21 draw the secant line $OQ_5$ through the origin with a slope 5 per cent less than the tangent OA to the initial part of the record. Draw a horizontal line representing a constant

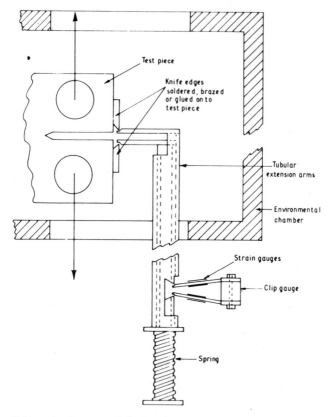

*Fig. 20 Schematic diagram of clip gauge extensometer.*

force of 0·8 $Q_Q$; $q_i$ is the distance along this line between the tangent OA and the force/displacement record. If this deviation from linearity at the force of 0·8 $Q_Q$ is greater than ¼ of the corresponding deviation, $q$, at a force of $Q_5$ excessive non-linearity is present and the curve is rejected.

Even when these conditions and the test piece thickness and crack length requirements were met certain materials still exhibited ostensibly valid $K_{Ic}$ values which varied with test piece dimensions (**6/6**). For this reason the current draft Standards stipulate that if the ratio $Q_{max}:Q_Q$ exceeds 1·10 then the $K_Q$ value is not equal to $K_{Ic}$, even if the other criteria are met (**6/7**).

There is also a (presently non-mandatory) recommendation that the test

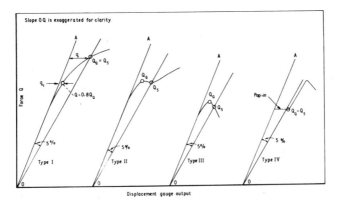

*Fig. 21 Principal types of force/displacement records showing quantities involved in analysis.*

piece crack length and thickness should be greater than $4 \cdot 0 \, (K_{Ic}/\sigma_Y)^2$ so that deviations from linearity are insignificant and $K_{Ic}$ values can be calculated directly from the maximum force.

6.2.1(*g*) *Calculation of* $K_{Ic}$. The stress intensity factor of the tip of a crack of length $2a$ in a uniformly stressed plate of infinite width is equal to $\sigma\sqrt{\pi a}$. In standard fracture toughness tests allowance is made for the influence of point loading and finite test piece dimensions on this relationship through the compliance function $Y$, such that

$$K_I = \frac{QY}{BW^{\frac{1}{2}}} \tag{5}$$

$Y$ values derived from the relationship between test piece compliance and dimensionless crack length $a/W$ (**6/8**) are given in tabular form for values of $a/W$ between 0·45 and 0·55 in increments of 0·001 $a/W$. Compliance functions of a wider variety of test pieces are also available for $a/W$ values between 0·30 and 0·70 (**6/9**).

6.2.1(*h*) *High rate* $K_{Ic}$ *tests.* $K_{Ic}$ values may increase, decrease or remain constant towards higher values of loading rate depending upon the behaviour of the material in question (**6/10**). In the main part of DD3 the loading rate in terms of the rate of increase in $K$ per unit time $K$, immediately prior to crack extension, is within the range 0·5 to 2·5 $MNm^{-\frac{3}{2}}$ per second. However information may also be required on the fracture behaviour of materials at higher loading rates. Suitable procedures will be incorporated in the proposed British Standards in the near future, based on the recommendations of the IIW Commission X UK Briefing Group—Dynamic Testing Sub-Group.

Conventional procedures are in fact suitable for $K$ rates up to about $10^4$ $MNm^{-\frac{3}{2}}s^{-1}$. In order to calculate $K$ from such tests it is necessary to obtain an additional recording of force versus time, or to make timing marks in the force/displacement records.

At $K$ values higher than about $10^4$ $MNm^{-\frac{3}{2}}s^{-1}$ conventional techniques are no longer suitable due to inertial effects in the displacement gauges. In addition autographic recorders are not fast enough and recording equipment such as cathode ray oscilloscopes must be used. A recommended alternative method is to record force as a function of time (**6/10**). This is suitable where the plastic deformation before fracture is insignificant and the maximum load recorded during the test is used directly to calculate $K_{Ic}$, Type III in Fig. 21. For this purpose the test piece size requirements are that the crack length $a$ shall be not less than 4·0 $(K_{Ic}/\sigma_Y)^2$.

In order to obtain linear load-time curves and a constant value of $K$, it is essential that the displacement rate, $V$, does not vary throughout the test. Servo-hydraulic machines are generally used in such tests. Kinetic energy (i.e. impact) machines are suitable for very high rate tests with $K$ of about $10^6$ $MNm^{-\frac{3}{2}}s^{-1}$. Problems arise in such tests due to inertial loads registered by the load transducers at the loading points (**6/11**). Because of this factor the upper limit to $K$ for dynamic tests using conventional techniques is about $10^5$ $MNm^{-\frac{3}{2}}s^{-1}$ which is within the range of some electrohydraulic test machines used in fracture mechanics laboratories.

$K_{Ic}$ values can be determined at higher rates by direct instrumentation of the test piece, the fracture force being given by the expression:

$$Q_f = \frac{Vt_f}{C_m + C_s} \tag{6}$$

where $t_f$ is the time to fracture the test piece and $C_m$ and $C_s$ are the dynamic compliance values of the test machine and test piece respectively (**6/10**). The time to fracture has been measured successfully using thermocouples and strain gauges attached to the crack tip and by observing changes in the force time curves from the transducer on the tup (6/12). Such techniques are specialized, however, and the difficulties involved in doing these tests (6/13) are likely to make standardisation difficult.

### 6.2.2 COD testing

Both COD and $K_{Ic}$ tests are performed using a unified testing procedure and therefore employ similar test pieces, fatigue pre-cracking procedures, testing procedures and analysis.

The crack opening displacement is a measure of the resistance of materials to fracture initiation under conditions where gross plastic deformation occurs and linear elastic fracture mechanics becomes invalid. The main objective of the COD test is to determine the critical crack opening displacement at the tip of a sharp crack at the onset of crack extension. This is done by measuring the displacement at the mouth of the notch using a gauge identical to that used in $K_{Ic}$ tests and by performing a suitable calculation.

6.2.2(a) *Test pieces.* Because there is no limitation to the amount of plastic deformation which is permissible the size of the test piece is not critical. However the thickness ($B$) of the standard three point bend test piece should be equal to that of the material under examination.

6.2.2(b) *Fatigue pre-cracking.* For COD tests the maximum fatigue stress intensity shall not exceed the ideal plane strain limit, i.e.

$$K_f < 0.63 \, \sigma_Y B^{\frac{1}{2}} \tag{7}$$

Similar conditions apply to the shape, length and curvature of the crack formed during fatigue for both $K_{Ic}$ and COD testing.

6.2.2.(c) *Testing.* The test record consists of a plot of force $Q$ versus crack opening displacement at the notch mouth $V_g$, measured by a displacement transducer.

A typical record is shown in Fig. 22, where the critical displacement $q_c$ at instability is the total value corresponding to the maximum applied force $Q_c$, Fig. 22(I). In certain cases test records similar to Fig. 22(II) are obtained. If the falling load portion can be shown to be associated with

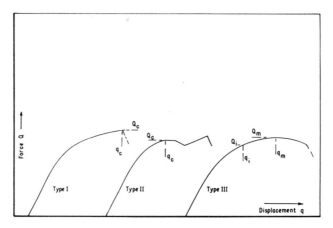

*Fig. 22  Force/displacement records for calculation of COD.*

crack growth, by the electrical potential method described below, or by an audible sound, then $Q_c$ is measured at the discontinuity shown.

The third type of record Fig. 22(III) where there is no sudden discontinuity is associated with slow stable crack growth which commences during the rising force portion of the test record. Since there is no visible or audible indication of ductile crack growth an alternative method such as the electrical potential method must be used to detect the value of COD at the initiation of stable crack growth $\delta_i$.

In those cases where crack growth cannot be measured the crack opening displacement $\delta_m$ at the displacement $q_m$ corresponding to the first attainment of a maximum load can be used.

6.2.2(d) *Interpretation.*  Several methods (**6/14**–**6/16**) have been propsed for converting $q_c$ values to $\delta_c$, two of which are used in the current Draft for Development.

All the methods assume that plastic deformation occurs by a hinge mechanism about a centre of rotation at a depth $r(W - a)$ below the crack tip.

The first method, due to Wells (6/14), is based on a theoretical approach. The equations for this method are:

$$\delta_c = \frac{0\cdot 45\,(W-a)}{0\cdot 45W + 0\cdot 55a + Z}\left[q_c - \frac{\gamma\sigma_Y W(1-v^2)}{E}\right] \quad (8)$$

for

$$q_c \geqslant \frac{2\gamma\sigma_Y W(1-v^2)}{E}$$

$$\delta_c = \frac{0\cdot 45\,(W-a)}{0\cdot 45\,W + 0\cdot 55a + Z}\cdot\frac{q_c^2 E}{4\gamma\sigma_Y W(-v^2)} \quad (9)$$

for

$$q_c < \frac{2\gamma\sigma_Y W(1-v^2)}{E}$$

where $\gamma$ is a non-dimensionalized limiting value of elastic clip gauge displacement.

Suitable tables are provided in the Standard to assist in the calculations.

The second method is based on experimental calibrations using test pieces up to 50 mm thick (6/15). For $\delta_c$ values between ∼0·0625 and 0·625 mm good approximations can be obtained using the expression:

$$\delta_c = \frac{(W-a)q_c}{W + 2a + 3z} \qquad 10$$

When the British Standard is brought out, however, the above expressions are likely to be replaced by a simplified version of the form:

$$\delta = \frac{K^2}{2\sigma_y E} \times \frac{0\cdot 4\,(W-a)\,q_p}{0\cdot 4 + 0\cdot 6a + z}$$

where $q_p$ is the plastic displacement equal to the total displacement $q$ minus the elastic displacement at a stress intensity factor $K$ calculated from the load and crack length using expression (5).

It should be noted that there is considerable controversy about the value of COD as a measure of fracture toughness in defect tolerance calculations (6/17). For instance $\delta_i$ values measured at the initiation of stable crack growth in the rising force portion of the test record are not necessarily related to $\delta_c$ values at instability, although in the Draft Standard no distinction is made between $\delta_i$ and $\delta_c$ values (in fact $\delta_i$ is not referred to). These omissions are likely to be rectified in the full British Standard where $\delta_i$, $\delta_c$ and $\delta_m$ will all be defined.

#### 6.2.2(e) *High rate COD tests.*
Conventional techniques may be used to measure $\delta_c$ up to rates limited by the performance of the clip gauge—about $0{\cdot}15$ ms$^{-1}$. An alternative method to measuring the displacement $q_c$ is to record force as a function of time in exactly the same way as for high rate $K_{Ic}$ tests **(6/18)**.

From a typical load-time curve, Fig. 23, the notch mouth opening displacement for a bend test piece is given by the relationship:

$$q_c = C_1 Q_c + \left[ \frac{a + 0{\cdot}45\,(W - a)}{W} \right] V t_\rho \qquad (11)$$

where $C_1$ is the compliance of the test piece at the notch mouth, $t_\rho$ is the time interval during plastic deformation of the test piece, Fig. 23.

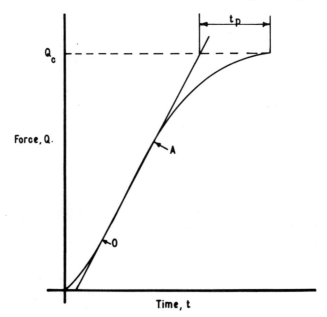

Fig. 23  *Force/time curve*

The values of $q_c$ so determined may be used to derive values of $\delta_c$ from expressions 8 and 9.

6.2.2(f) *Crack monitoring.* The electrical resistance technique enables a direct measurement of crack length to be obtained and is simple to use and relatively inexpensive. This method is based on measuring the voltage drop across the notch of the test piece **(6/19)**. The essential components of the apparatus required are:

(1) A stable DC current source capable of supplying $10A$ per $cm^2$ of test piece cross section.
(2) Copper current leads screwed or preferably soldered onto the test piece.
(3) Electrical potential leads spot welded onto the test piece.
(4) A recorder capable of measuring voltages down to $10\ \mu v/cm$.

Crack lengths are determined from a calibrated plot of $a/W$ versus voltage. In COD tests the potential across the test piece may be plotted against the output from the clip gauge, Fig. 24.

The inflection at I is associated with the first formation of significant plastic deformation at the crack tip **(6/20)**. The initiation of ductile crack growth occurs at inflexion $A$ and fibrous crack growth occurs at $B$.

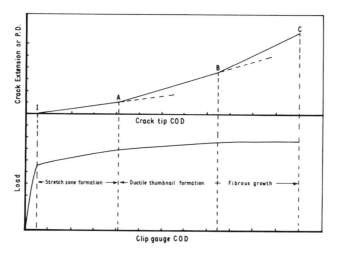

*Fig. 24 Crack extension during COD testing.*

## 6.3 NON-STANDARD METHODS

### 6.3.1 *R* curve analysis

The theory of crack growth resistance ($R$) curves has been dealt with elsewhere in this book **(6/21)**, the basic requirement being that some measure of the fracture resistance (e.g. $G$ or $K$) is determined as a function of the change in crack length of the test piece.

A testing procedure which has been used in the United States **(6/22)** utilizes a compact tension type of test piece. The test piece is restrained from buckling by fixing rigid face plates to the test pieces in critical regions. The crack opening load is applied by driving a wedge between tapered blocks and circular segments on either side of the test piece notch. The function of the wedge is to provide a very stiff system to prevent fast fracture.

Crack growth is monitored by the use of a double compliance system where crack opening displacement measurements are taken at two separate $a/W$ positions on either side of the load line. The load is calculated from a knowledge of the crack opening displacement and the compliance of the test piece. The crack length is determined from the calibrated ratio of the two displacement gauges as a function of $a/W$.

However, instead of using wedge opening to provide a stiff system it is possible to use servo-controlled electrohydraulic machines controlled from the output of one of the displacement gauges and the electrical potential technique can be used to monitor crack growth.

An alternative approach to R curve determination introduced by the Naval Research Laboratories **(6/23)** has much to recommend it from the point of view of simplicity. Test pieces of different thickness $B$ and width $W$ are fractured in impact and the total energy to fracture, $U$, recorded. A plot is constructed of the fracture energy divided by the ligament area of the test pieces, $U/B(W-a)$ against the ligament length, $W-a$. The NRL workers proposed that the fracture energy was given by

$$U = R\rho(W-a)^2 B^{0.5} \tag{12}$$

where $R\rho$ is a constant associated with the inherant resistance to fracture of the material. More recent work, however, **(6/24)** has shown that the total energy to fracture is given by the expression

$$U = R_c B(W-a) + S_c B(W-a)^2 \tag{13}$$

where $R_c$ is a measure of the fracture resistance of the material possibly related to $G_{Ic}$ and $S_c$ is an energy density value governing the resistance to plastic deformation. This expression is in line with the observation that in fact the total fracture propagation energy decreases with increasing crack extention in test pieces of finite size.

### 6.3.2 J integral

$J_{Ic}$ values are currently determined from a series of tests in which the crack growth is monitored (6/25). One technique used (6/26) is as follows:
(a) Load the test piece to different displacement values with a test machine in displacement control.
(b) Unload and mark the extent of crack growth using a heat tinting procedure (180°C for ten minutes).
(c) Fracture each test piece at a low temperature (e.g. $-150°C$) and measure the crack extension.
(d) Calculate $J$ values from the load versus loadline displacement record using the relationship:

$$J = \frac{2U}{B(W-a)} \tag{14}$$

where $U$ is the area under the load displacement curve at the displacement of interest.
(e) Plot a curve of $J$ versus crack extension.
(f) Fit a straight line $J = 2\sigma_f \Delta a$ to the best fit line drawn through the $J$ versus $\Delta a$ points, where $\sigma_f = (\sigma_Y + \sigma_u)/2$ and $\sigma_u =$ Ultimate Tensile Strength. The critical value of $J$ is that which occurs at the intersection of these two lines, Fig. 25.

A specially designed test piece (Fig. 25a) is employed to enable loadline displacements to be measured so that the area under the load displacement curve is equal to the fracture energy. It is possible to construct the $J$ versus $\Delta a$ curve shown in Fig. 25 from a single test by using the electrical potential technique (6/27).

### 6.3.3 The equivalent energy procedure

A recent method for predicting $K_{Ic}$ values from small scale non-valid tests is the equivalent energy (6/28) concept. The same techniques are used as for the $J$ integral procedure above in which a load deflection curve is constructed for the determination of energy values.

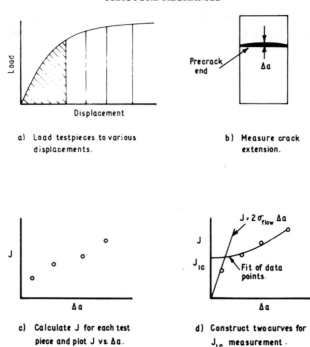

Fig. 25   Procedure for $J_{Ic}$ measurement.

The procedure can be summarized as follows (6/29):

1 Select any load on the linear porton of the load-deflection curve ($Q_B$). Measure the area under the load deflection curve up to maximum load and divide this by the area up to $Q_B$. The ratio of the two areas is $d$.
2 Calculate a value of stress intensity as follows:

$$K = \frac{Q_B \, Y d^{\frac{1}{2}}}{BW^{\frac{1}{2}}} \tag{15}$$

Buchalet and Mager (6/27) found that values of stress intensity measured in this manner on 1 in thick compact tension test pieces were in agreement with $K_{Ic}$ values obtained from valid $K_{Ic}$ tests on 12 in thick test pieces at temperatures up to about 10°C. However, at higher temperatures a systematic increase in $K_{Ic}$ values was observed with increasing test piece thickness.

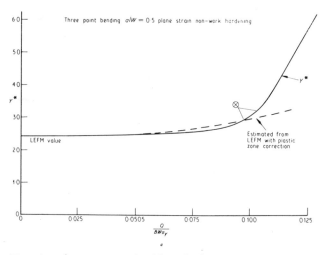

Fig. 25a. *Test piece for measuring load line displacement.*

### 6.3.4 Impact fracture propagation energy

With the advent of fracture mechanics testing, it was realized that by using impact test pieces pre-cracked in fatigue the fracture propagation energy could be measured directly. This energy is divided by the fracture area to give a $U/B(W-a)$ value which may be related to the fracture toughness, $G_{Ic}$, of the test piece **(6/30, 6/24)**.

Test machines with impact energy capacities of less than 50 Joules and capable of measuring to within an accuracy of 0·01 Joules are recommended for this type of work. Windage losses due to the friction of the test machine and of the air are accounted for by prior calibration. In addition, with low strength materials energy may be absorbed in indentation of the test piece prior to fracture.

Finally kinetic energy may be absorbed in accelerating the broken halves of the test piece out of the anvil. This can be determined by measuring the energy absorbed by the broken halves of a test piece held together by a weak adhesive.

$U/B(W-a)$ values are sometimes measured in instrumented tests by calculating the energy absorbed from the area under a load-displacement or a load-time curve. In this case windage and kinetic energy losses do not have to be accounted for.

A Draft Standard (**6/31**) has been written for pre-cracked Charpy tests but this has only recently been circulated for comment.

## 6.4  ACKNOWLEDGEMENT

The author wishes to thank Dr K. J. Irvine, Manager, Sheffield Laboratories, British Steel Corporation, for permission to publish this paper.

# 7

## Application of Fracture Mechanics to the Brittle Fracture of Structural Steels

R. R. BARR
*British Steel Corporation, Motherwell, Lanarkshire*

P. TERRY
*British Steel Corporation, Motherwell, Lanarkshire*

### 7.1 INTRODUCTION

In any structure, tension stresses inevitably occur in some regions and brittle failures can initiate from defects located in these areas, especially when there are associated stress concentration features and residual stresses due to welding. In practice the significant defects are usually located in a weld region and in addition to the effects of stress concentration and residual stress, the defects may be associated with areas subjected to adverse metallurgical changes. Fractures, once initiated, can propagate rapidly under low applied stresses, causing extensive damage. Brittle fractures have been observed in the past in bridges (**7/1**), pressure vessels (**7/2**), storage tanks (**7/3**), pipe lines (**7/4**), and the, now classical, welded ships (**7/5**). Major catastrophic brittle fractures are relatively rare today, despite a move to more onerous design conditions, and this is due to research carried out to gain an understanding of the many parameters which have to be controlled to eliminate the possibility of fracture.

Initially, this research concentrated on use of Charpy vee notch impact tests, but many other laboratory tests such as the Robertson crack arrest test (**7/6**), the Pellini drop weight test (**7/7**) and the Wells wide plate test (**7/8**), were devised in an attempt to obtain design information on brittle

*The MS of this paper was received at the Institution of Mechanical Engineers on 9 June 1975 and accepted for publication on 22 August 1975.*

fracture resistance. In their original form, the aim of these tests is to characterize material by means of the brittle/ductile transition behaviour, the intention being to use materials above the brittle/ductile transition temperature to ensure freedom from brittle fracture.

More recently in the United Kingdom, design of support and containment structures (excluding pipelines) has been based on the avoidance of fracture initiation rather than the more severe criterion of design on the basis of resistance to fracture propagation. In this context, the Wells wide plate test, resembling a particular service condition was the first test to successfully demonstrate the conditions for large scale brittle fracture in the laboratory. This test was devised to reproduce the conditions involved in the fracture of a large welded oil storage tank in 1952 **(7/9)** where failure initiated from small defects which had been subjected to strain age embrittlement in the sub-critical heat affected zone. In order to examine this phenomenon Wells used a test plate, developed from a test devised by Greene **(7/10)**, containing notches sawn into the weld preparation prior to welding. The large test plate (915 mm × 915 mm) was used to reproduce service conditions by retaining high residual stress levels within the plate. Using this test, it was demonstrated that low-stress brittle fractures could be obtained at temperatures, until then, considered safe for such materials.

The accumulated results of wide plate tests have been used to define safe minimum operating temperatures for steels, based on the temperature for 0·5 per cent strain **(7/11)** (subsequently changed to 4 × yield point strain, $e_Y$) in a wide plate test, representing the most severe strain conditions in the nozzle area of a pressure vessel **(7/12)**. Using such results design standards **(7/13–7/15)** have been drawn up for C-Mn steels giving minimum design temperatures against the corresponding plate quality expressed as the temperature to achieve a given Charpy energy (the material reference temperature, MRT), as shown in Fig. 26, an extract from BS1515 Appendix C. The MRT is the temperature for 27 or 41J impact energy for UTS less than or greater than 450 N/mm² respectively.

Some time previous to this, the original concept of fracture mechanics had been formulated by Griffiths **(7/16)** on the basis of an energy balance criterion. Irwin **(7/17)** subsequently developed the application of this early theory to the fracture of metals introducing the concept of stress intensity within the framework of what is now known as linear elastic fracture mechanics (LEFM). The basic concept of LEFM has limited application to structural steels since the theory only applies to essentially elastic fracture

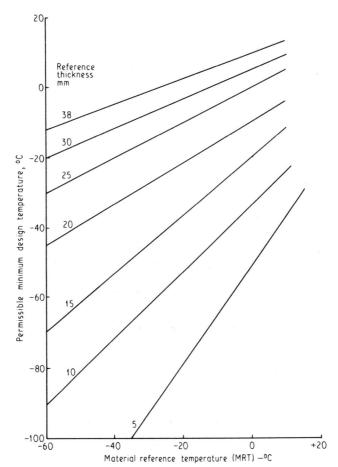

*Fig. 26  Extract from BS1515 Appendix C for as-welded components.*

and in the majority of structural steels, even apparently brittle, fractures are accompanied by more plasticity than the theory will permit. To cover these situations the basic theory was later extended by Wells (**7/18**) and Cottrell (**7/19**) to the elastic-plastic régime involving the concept of crack opening displacement (COD).

In connection with the development of the COD approach, the results from large scale wide plate tests have been used to study the application of small scale COD tests to the behaviour of large components leading to a procedure for the derivation of a relationship between the COD measured in a small scale laboratory test and the defect tolerance of a structure.

The current approach to the brittle fracture assessment of structural steels is now based upon small scale fracture toughness testing combined with wide plate tests specifically designed to examine particular service situations in terms of tolerable defect sizes.

## 7.2 THE CONCEPT OF TOLERABLE DEFECTS

In engineering terms, the development of fracture mechanics, as a means of defining quantitatively the critical parameters which control fracture initiation, has provided the most significant advance in controlling the problem of brittle fracture. A number of fracture toughness test measurements can be made as indicated in other papers in this series and these include $K$, $G$, $J$, $\delta$ and $R$. In practical terms the most important of these have proved to be $K_{1c}$ the critical stress intensity factor for opening mode failure and the extension to more ductile fractures using COD. The use of these parameters in defect control relies upon locating and sizing defects in a structure, combined with a knowledge of the stresses acting upon those defects.

The $K_{Ic}$ parameter is calculated in the laboratory from acting stresses and specimen geometry. The basic equation for the calculation of $K_{1c}$ is:

$$K_{1c} = \sigma_c f(a) \tag{1}$$

where for a test specimen $\sigma_c$ is the fracture stress, $a$, the crack length and the function contains factors to allow for specimen size, shape and loading system. In its simplest form, that of a through-crack in an infinite plate, the function can be represented by $\sqrt{(\pi a)}$. Hence

$$K_{Ic} = \sigma\sqrt{(\pi a)} \tag{2}*$$

Using the laboratory determined $K_{Ic}$ value and a knowledge of the stresses acting in a component this simple equation can be rearranged to give values of defect size $a_c$ which will just cause failures in the component, i.e.

$$a_c = \frac{1}{\pi} \left(\frac{K_{Ic}}{\sigma}\right)^2 \tag{3}$$

* See footnote to General notation on page vii.

In practical components the function of '$a$' is more complex and contains factors to permit analysis of different defect shapes and locations **(7/20)**.

For engineering applications using general structural steels the main use of fracture mechanics lies beyond the limits of linear elasticity and there are two methods of analysis. The first of these extends the stress intensity approach by means of plasticity correction factors **(7/21)** while the second, and perhaps the more widely used, relies on the crack opening displacement as a suitable parameter.

The theoretical equation of COD again relates toughness with applied stress and crack size, as in the case of $K_{Ic}$, and the COD is given by **(7/22)**

$$\delta = \frac{8\sigma_y a}{E} \log_e \sec\left(\frac{\pi\sigma}{2\sigma_y}\right) \tag{4}$$

Up to the limit $\sigma/\sigma_Y \leqslant 1$ this equation is directly relatable to the equations of linear elastic fracture toughness $K_{Ic}$, and $K_{Ic}$ and $\delta$ can be shown to be related by the equation

$$\delta = \frac{K_{Ic}^2}{\sigma_y E} \tag{5}$$

At stresses below $\sigma/\sigma_Y = 1$, therefore defect sizes can be calculated which will cause failure in a component having given COD and strength properties.

Above the yield stress, when the theoretical analysis limits are violated, it has been shown by Wells **(7/18)** and Burdekin and Stone **(7/22)**, that the relationship between COD and overall applied strain is linear. In order to obtain a unified approach to defect tolerance calculations Burdekin and Dawes **(7/23)** produced a design line based upon experimental observations of COD versus applied strain in wide plate tests. The most recent version of this design line **(7/24)** shown in Fig. 27, is compatible with previous theoretical and experimental work. To make safe predictions of acceptable defect size, the design line has been drawn as the conservative limit of available experimental data of observed COD versus strain in large scale tests and it has been estimated that the tolerable defect size calculated, $a$, using the design line, is approximately half the size likely to cause failure The design line is plotted as a function of the dimensionless COD parameter $\phi = \delta/(2\pi e_Y a)$ against dimensionless strain $e/e_Y$ where $e$ is applied strain and $e_Y$ the yield strain. Thus for a given component if the level of

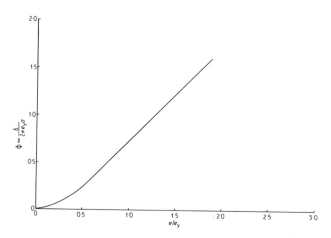

*Fig. 27  Non-dimensional COD vs. non-dimensional strain* (**23**).

applied strain in the region of a defect is known, then a value for $\phi$ can be obtained from Fig. 27 and a defect tolerance calculated from:

$$a = \frac{1}{2\pi\phi} \left(\frac{\delta}{e_y}\right) \tag{6}$$

For ease of manipulation the curve has been replotted in terms of $1/(2\pi\phi)$ versus $e/e_Y$ and the term $1/(2\pi\phi)$ is given the symbol $C$ (see Fig. 28). The value of $C$ can be calculated from:

$$C = \frac{1}{2\pi \left(\dfrac{\sigma}{\sigma_y}\right)^2} \text{ for } \frac{\sigma}{\sigma_y} \leqslant 0\cdot 5 \tag{7}$$

or

$$C = \frac{1}{2\pi \left(\dfrac{e}{e_y} - 0\cdot 25\right)} \text{ for } \frac{e}{e_y} \geqslant 0\cdot 5 \tag{8}$$

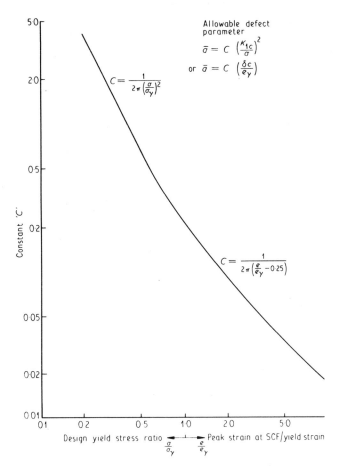

*Fig. 28  Values of constant 'C' for different loading conditions.*

The values of $\sigma$ or $e$ must take into account all the stresses acting on the defect and should include the applied stress, the effect of stress concentrations and the effect of residual stress caused during fabrication, i.e.

$$\sigma = \sigma_a \times SCF + \sigma_R \tag{9}$$

where $\sigma$ is the stress acting on the defect

$\sigma_a$ is the applied stress

$\sigma_R$ is the residual stress

At stress values above $\sigma/\sigma_y > 1$, the concept of applied strain is currently used and stresses and strain are simply related via $e = \sigma/E$
Using the C notation equation 6 can be written

$$a = C \left(\frac{\delta}{e_y}\right) \qquad (10)$$

Since $\delta$ and $K_{Ic}$ are directly related up to $\sigma/\sigma_y \simeq 1$, by the relationship $K_{Ic}^2 = E\sigma_y\delta$ this equation can also be written:

$$a = C \left(\frac{K_{1c}}{\sigma_y}\right)^2 \qquad (11)$$

giving a unified defect tolerance calculation.

In inspecting actual components the size, shape and orientation of defects revealed by non-destructive testing will vary and in order to compare the sizes of actual defects with the calculated maximum allowable size it is necessary to convert the actual defects to those of idealized surface, embedded or through-thickness cracks. Procedures for this are available and they are included along with a defect analysis procedure, in a recent paper by Burdekin (**7/25**).

These procedures, as already stated, give safe defect tolerance levels and in a number of cases it is necessary to determine the degree of conservatism between the calculated tolerable defect size and the sizes of defects to cause failure, under specific service conditions, the critical defect size $a_c$. To this end, wide plate tests have been designed to assess the significance of various defect types (for example surface and through thickness) located in the different regions of a welded joint for example, weld metal and heat affected zone. These tests can give an accurate picture of the combined influence on critical crack size of applied stress and residual stress, and can also be used to assess the influence of stress concentration features by placing defects close to welded attachments. The results of such tests performed by a number of workers (**7/26, 7/27**) have confirmed that the defect tolerance calculations produce conservative predictions of $\bar{a}$ with reference to $a_c$. In addition these large scale test results also indicate the limitations of the analysis at temperatures where materials are more ductile, the estimate of $\bar{a}$ being extremely conservative. It has recently been shown **(7/28)** that the defect tolerance calculation becomes invalid when small scale laboratory test specimens exhibit fully ductile behaviour. In these circumstances there is no longer a fracture problem and a plastic collapse criterion provides a more meaningful guide to performance.

In actual service conditions the problem of brittle fracture cannot be considered in isolation and other failure modes must be considered including fatigue and stress corrosion, which can cause growth of small defects towards the critical crack size, and the methods of determining the effects of such mechanisms are outlined in later papers in this series.

## 7.3 ENGINEERING FRACTURE MECHANICS TESTS

In a preceding paper, the standard test methods for fracture toughness testing have been outlined, however, as in all applications related research the gathering of practical information is much more difficult. This section outlines the procedures and techniques necessary to obtain practical test data, and is essentially confirmed to COD and wide plate testing as these particular tests are more normally applied to the grades of steel used in the majority of engineering constructions.

In view of the significant effect of thickness on the brittle/ductile transition and also to some extent on the levels of COD obtained, there is a basic requirement that the tests be performed on specimens of a thickness equal to that used in the fabrication of the structure under consideration, in order to achieve equivalent constraints in both situations. It is essential therefore that data be acquired over the range of thickness at which the steel is supplied for service. Furthermore, it is necessary to test these steels over a range of temperatures, appropriate to various minimum service temperatures. Loading rates must also correspond to those encountered in service, although pseudostatic tests using slow rates of loading represent a large number of applications. It is also important to consider the effect of stored energy in the testing arrangement and its influence on whether crack growth is stable or unstable. This stored energy should approximate to that expected in the structural detail under consideration. Test specimens are normally fatigue pre-cracked to simulate a sharp defect, although in some instances naturally occurring weld metal cracks and other defects can be used. However, the requirement for fatigue pre-cracking of specimens can itself produce difficulties, in obtained straight fatigue crack fronts **(7/29)**.

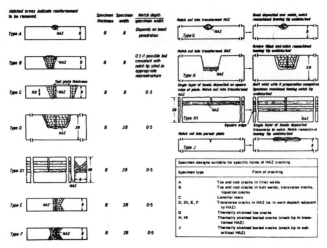

*Fig. 29    Specimen designs for HAZ toughness testing (after Dolby (7/30)).*

After consideration of these testing requirements the next step is to acquire data relevant to engineering structures and this, in all but a few specialized cases, involves assessment of welded joints. In the main, parent steel properties are superior to those obtained in the heat affected zone of a weld or in the weld metal and at normal operating temperatures and thicknesses, the plate toughness properties are usually so high that brittle fracture initiation is not a problem. In assessing the weld metal and heat affected zone using COD test speicmens, it is important to take into account the effects of defect location and orientation. To assess these different features a number of specimen designs have been developed (7/30) and some of these are illustrated in Fig. 29 (7/31) for heat affected zones.

General assessment of weld metal, for example, can involve specimens in which the crack path runs along the weld centre line normal to the plate surface and also specimens in which the crack path can travel from the weld surface through the thickness of the plate since these two orientations may give significant differences in toughness level and are representative of two modes in which a natural crack can occur. In the testing of heat affected zones the situation is more complex. To determine the toughness of a heat affected zone, welded joints can be prepared using $K$ weld preparations (Fig. 29 type E) in order to allow the notched laboratory test to sample a greater proportion of the HAZ, however, such weld preparations are not common in normal constructions and double or unequal $V$ preparations are much more widely used. In testing the latter preparations with a specimen orientation in which the crack runs normal to the plate surface this necessitates a notch which crosses both weld metal, heat affected zone and in some cases parent plate material and it is extremely difficult to analyse the resulting test information. It has also been shown that the testing of weldments and the assessment of heat affected zones in weld metal can be further complicated by the direction of the crack inasmuch as different results can be obtained in testing, for example, a heat affected zone if the crack passes from weld metal into the heat affected zone and in the case where the crack passes from parent plate into the heat affected zone (**7/31**). This difference is caused by the different plastic zone shapes which generate at the tip of the crack between weld metal and heat affected zone and parent plate.

In addition, testing must be considered on weldments over the range of heat inputs, welding consumables and welding processes which can be applied to a particular steel in practice, and should be carried out on material in the same heat treated condition as will be used in service.

The preceding list of factors which have to be considered appears initially to be formidable but in specific applications it is possible to restrict the number of variables and to design a laboratory test programme either to determine the toughness requirements or, if these are agreed, to select material and weld procedures which will meet these requirements. In addition to the small scale tests, it may be necessary to perform wide plate tests specially designed and notched in the area judged to be most critical. These wide plate tests will give some indication of the levels of conservatism in the defect tolerance analysis. However, in order to be able to apply the results in a less conservative manner to actual structures a great deal of information is required concerning the expected design loads in service and the effect of the different estimates of these design loads are detailed in the next section.

## 7.4 COMPLEMENTARY DESIGN AND SERVICE INFORMATION

Before fracture toughness measurements taken from laboratory tests, $K_{Ic}$ or COD, can be put to practical use they have to be translated to a defect size, as explained in a previous section. This defect size may be termed either as critical or tolerable. To perform this calculation it is necessary to have available a measure or estimate of the stresses involved in the particular part of the structure which is being analysed. The stresses are either used directly as denoted in equation (7) or converted to an equivalent strain value as used in equation (8) to characterize the conditions which exist at the tip of the pre-existing defect.

If accurate stress values are not available, the estimates made should be conservative so that the defect size calculated, errs on the safe side. Values for the nominal stresses in a particular structure are usually available since they form part of the general overall design calculation. It is the estimates for residual stress, stress concentration and the way in which they are used which can cause difficulties. Stress concentrations can be estimated by a variety of procedures including photoelastic techniques and the essentially mathematical finite element analysis method, whilst more direct values can be obtained by 'in-service' strain gauging. It is important to realise that the stress concentration influence only extends over a short range, from the stress concentration feature of the joint. An example of this is illustrated in Fig. 30 which summarizes the stress concentration factors at the toe of 45° fillet welds of different sizes relative to plate thickness. This data has been published in a paper by Barr and Burdekin (7/32) from original finite element computations by Hayes and Maddox (7/33). The calculations show that the magnification effect at the toe of the weld decays very rapidly and even for a fillet of leg length equal to the plate thickness the effect has largely diminished to unity at a distance of 0·1 times plate thickness from the stress concentration feature.

In a number of current documents (7/34) it is considered that the stress concentrations should be regarded as existing over a uniform field up to a distance of 0·15 times plate thickness from the stress concentration feature. The paper by Barr and Burdekin suggests that appropriate values of the stress concentration factor $(M_K)$ may be selected from Fig. 30 or other figures drawn to represent different geometric details. This value can then be applied using the normal formula for defect tolerance to derive the value of $C$ as shown previously.

Residual stresses due to welding have also to be taken into account as in section 7.2. The maximum residual stress which can exist corresponds to the yield strength of the material and on some occasions the value assumed can have a pronounced effect on the defect size calculated.

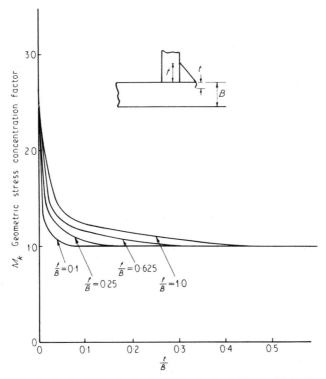

*Fig. 30  Geometric stress concentration factors due to fillet weld (45°).*

This occurs where the service stresses are low, reducing the effectiveness of the term $\sigma_a \times$ SCF in equation (9). This situation is best illustrated by a specific example, concerning works plant fabricated in 100 mm thick mild steel with a maximum operating stress of 45–75 N/mm². Defects were found in the weld during routine inspection caused by lack of fusion. The toughness of the weld metal in which the defects were located was 0·074 and 0·091 mm COD as determined from laboratory toughness tests, in the as-welded and stress relieved conditions respectively. A comparison of defect sizes calculated for the as-welded and stress relieved condition are shown in Table 1.

Comparing the tolerable defect sizes for conditions 1 and 2, indicates the small influence of the factor assumed for stress concentration where

*Table 1*

| Condition | $\sigma_a$ | SCF | $\sigma_R$ | $\bar{a}$ |
|---|---|---|---|---|
| 1 As-welded | 75 N/mm² | 1 | $\sigma_y$ | 6·3 mm |
| 2 As-welded | 75 N/mm² | 3 | $\sigma_y$ | 4·5 mm |
| 3 Stress relieved | 75 N/mm² | 3 | 0·25 $\sigma_y$ | 13·5 mm |
| 4 Stress relieved | 75 N/mm² | 3 | Nil | 29·0 mm |

the operating stresses are low. In addition, the calculation illustrates the marked influence of the value selected to represent residual stresses. There is some controversy about the effectiveness of thermal stress relief in reducing or eliminating residual stress. In the example given there is a significant difference in $\bar{a}$ value depending on the value assumed for residual stress (conditions 3 and 4).

It is also of value to examine in more detail the extent to which stress concentration features and residual stresses can influence the calculation of tolerable defect size, $\bar{a}$. Figure 31 illustrates, for a particular design stress level, how the assumed value for stress concentration can influence the defect tolerance calculated over a range of COD values from 0·1 to 0·4 mm. Figure 31 includes as-welded and stress relieved conditions, assuming in the stress relieved case the complete removal of residual stress and no changes in yield strength or measured toughness value. In practice changes in yield strength and toughness may in fact occur. The graph illustrates the pronounced improvement in defect tolerance after a stress relief, particularly where stress concentration factors are low. Also shown is the overriding effect of large stress concentration factors where differences between as-welded and stress relieved conditions at the same COD level are extremely small and that even a 4 times increase in COD level produces only small changes in defect tolerance. For the purpose of these calculations the stress concentration effect has been assumed to act over the whole defect length.

However, with regard to the influence of stress concentration and the distance over which it may be considered to operate, it is possible to construct a graph, as in Fig. 32, where the curves of defect tolerance versus stress concentration factor for two COD levels and the curve for the stress concentration factor versus distance for a tee butt weld are superimposed.

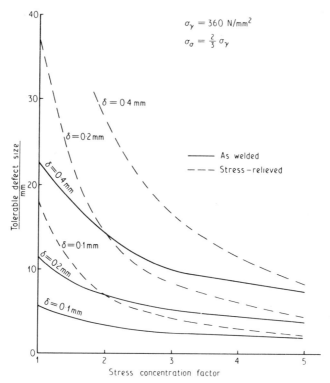

*Fig. 31  Influence of stress concentration on tolerable defect size.*

For a COD level of 0·05 mm, defects of tolerable depth can exist within the range of the stress concentration effects producing a situation where brittle fracture is likely from small defects located at the toe of a weld. On the other hand, for a COD level of 0·2 mm, the defect depth at any particular stress concentration factor level is always in fact greater than the distance over which that stress concentration operates which means that the actual defect tolerance is in reality greater than calculated.

A further factor reducing the possibility of fracture initiation in this latter case, is that the zone of low toughness associated with the weld is itself small and with increasing size the defect tip will have moved into an area of higher toughness.

A useful working diagram, relating applied strain, COD and yield

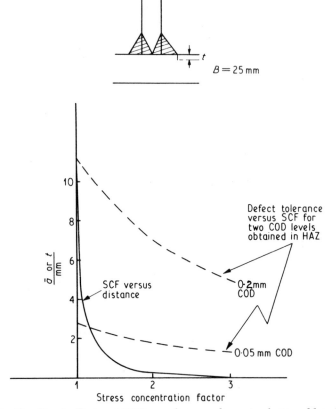

*Fig. 32  Combined effects of SCF vs. distance for a tee-butt weld and COD vs. SCF.*

strength to defect tolerance, as given by equation (10) is shown in Fig. 33. An illustration is shown on the diagram for a situation where the total applied strain is equal to $1.5 \times e_y$, the material toughness is 0.2 mm COD, and the yield strength is 400 N/mm². The defect tolerance, $\bar{a}$, is 12.5 mm. This diagram also illustrates the influence of the value assumed for yield stress.

Thus, it is possible to determine tolerable defect sizes from either $K_{Ic}$ or COD values as illustrated above, from suitably designed laboratory test pieces. In view of reservations regarding the conservatism which may be

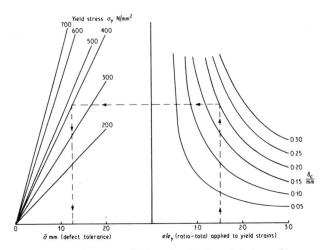

*Fig. 33  Nomogram relating applied strain, COD and defect tolerance.*

involved in the calculation, it may be necessary in some circumstances to use the more definitive wide plate test to determine fitness for service. In any event the wide plate test should be fully utilized in the preparation of design standards as illustrated below.

## 7.5  APPLICATION OF FRACTURE MECHANICS TO DESIGN AND MATERIAL SELECTION

Although fracture mechanics as illustrated above, provides a means of defining the critical parameters involved in unstable fracture initiation, there are, a number of aspects which require further research, before the achievement of an agreed rational basis for design which recognizes the existence of defects. In the United Kingdom the selection of steels for containment and support structures is based, by implication only, on the general assumption that undetected weld defects may be present and that the toughness of the material should be sufficiently high to avoid fracture initiation from these defects. However, there is as yet no uniformly agreed technical basis for a definition of the fracture toughness requirements applicable over a wide range of structures. It is obvious that the toughness requirements will change for various structures depending upon a number of factors as follows:

1. design or operating temperature;
2. nominal design stress;
3. stress concentration features;
4. applicability or otherwise of using thermal stress relief;
5. effect of proof loading;
6. consequences and risk of failure in terms of economics and danger to life.

At the present time the various material selection Standards are completely unrelated in terms of clear cut adjustments to the toughness requirements taking into account the features noted above. Containment standards such as BS 1515 Appendix C for pressure vessels, and BS 2654 and BS 4741 for storage tanks have used wide plate test results to assess fitness for service, the results of these tests being related to the Charpy impact test energy values for parent plate in the form of test temperatures to achieve certain minimum impact energies as shown by the example given in Fig. 26. Support structure standards such as BS 153 for Bridges and BS 449 for steel buildings quote, at present, a maximum thickness for a particular grade of steel, the level being mainly related to experience based on Charpy impact levels in parent plate. Thus the basis for selection is different and as a result structural standards permit some plate steels to be used up to 100 mm thick in the as-welded condition, whereas containment standards recommend stress relief above 38 mm for all grades covered by the standards. Comparison between these various standards is given in Fig. 34 for a specific commonly used design temperature. Although it may be assumed that these variations are due to differences in complexity of design it should be noted that modern bridges can have areas of stress concentration as high as, for example, nozzles in a pressure vessel, and some caution may be necessary in using steels up to the maximum thickness permitted for critical parts of these fabrications. Although, in theory, fracture mechanics analysis should be capable of resolving the different toughness levels which are required, there are a number of difficulties in the precise interpretation of some fracture toughness mea-

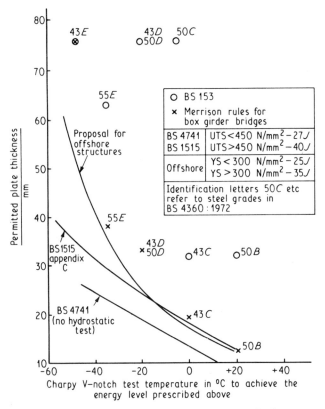

*Fig. 34  Comparison of toughness for various British Standards.*

surements for heat affected zones, weld metals, as indicated previously. Consequently, although the theory is fairly advanced, there is, as yet, no agreed procedure for a completely rational basis for design over a range of structural applications. An additional factor which prevents such a procedure is the lack of laboratory toughness data covering a sufficient range of materials including weld metal and heat affected zones. Thus, there is considerable scope for more authoritative data for a variety of steels over a range of section thickness and welding parameters.

Until interpretational difficulties associated with the COD test have been clarified, therefore, it would be inappropriate to use this test, on its own, to specify material toughness levels for national standards. However, at the present time there is sufficient experience in the use of the techniques for fabricators and users to make a safe decision on whether to rectify a defect detected either during fabrication or service and also, possibly, to form the basis of an internally agreed toughness specification for specific applications, based upon the detection levels for defects using non-destructive testing. In the present circumstances is it advisable to carry out wide plate tests designed to simulate practical welds, thus bridging the gap between the small scale laboratory test and the final structure. Reliance on COD testing alone could mean an over-conservative prediction of the toughness requirements with subsequent difficulties in material procurement or economic penalties in terms of prolonged fabrication times. It may also be possible to consider for many constructions a dual approach to toughness requirements. Thus for high stress concentration regions a different toughness requirement could be stipulated relative to regions stressed only to normal design stresses.

The Charpy vee notch impact test is currently used for quality control purposes both in the supply of steel to a given specification and for monitoring weld procedures during fabrication. Although there is no reliable theoretical correlation between Charpy energy absorption and fracture toughness, as measured by $K_{Ic}$ or COD, it is possible to select a Charpy level, which is consistent with given fracture toughness properties, as a minimum quality level. The Charpy test will continue to be used in this way because of its low cost and simplicity.

## 7.6 CONCLUSIONS

For most structures control of brittle fracture involves design against fracture initiation and in the UK test procedures are available to determine the factors controlling initiation. These tests involve the use of the theory of linear elastic fracture mechanics and its extension into the area of general yielding fracture mechanics.

The paper illustrates the calculation of defect sizes based on current theory, from measurements of $K_{Ic}$ or COD in laboratory tests and demonstrates the influence of various design factors, stress concentration and residual stress effects on the defect tolerance calculated.

It is pointed out that these calculations are safe but in some cases they may be over-conservative and in these circumstances the use of specially designed wide plate tests is recommended to determine the toughness levels required to give satisfactory service performance.

Since fracture initiation is likely to occur from weldments special attention should be paid to the design of laboratory fracture toughness tests to determine the fracture toughness of weld metals and heat affected zones. Some of the problems associated with the interpretation of the critical toughness value are discussed and it is concluded that special care is required to select a notch position and orientation appropriate to the types of defect which are likely to be encountered in practice.

Present procedures can form the bases for agreement between directly involved parties to decide whether to repair a defect already present, or to assist in the selection of minimum toughness requirements for a specific application based upon NDT discrimination levels.

Rationalization of existing standards to the specification of toughness requirements which are consistent with the present theories related to the various design factors requires additional data covering a wide range of steels and weld procedures.

It is considered that the Charpy vee notch impact test will continue to be used for quality control purposes, although the impact energy levels will be based upon the results of fracture toughness tests.

# 8
# Analysis and Application of Fatigue Crack Growth Data

## L. P. POOK
*Department of Industry, National Engineering laboratory, East Kilbride, Glasgow*

The phenomenon of metal fatigue has been studied for a long time, but it has only relatively recently been appreciated that most structures, particularly welded joints, contain crack-like flaws, so that virtually the whole fatigue life is occupied by fatigue crack growth. The fracture mechanics concept of stress intensity factor has proved particularly convenient for the analysis of fatigue crack growth data in a form which can be applied directly to engineering problems, and its use has led to a much better understanding of the fatigue behaviour of structures. The effects of interaction between different load levels are not yet fully understood; this is not necessarily a serious drawback as servo-hydraulic fatigue testing equipment permits the application of virtually any load history to structures or fatigue crack growth specimens.

**Notation**

*From standard list*

$a$
$E$
$K_I$
$R$
$Y$
$\sigma$
$\sigma_Y$

*The MS of this paper was received at the Institution of Mechanical Engineers on 9 June 1975 and accepted for publication on 22 August 1975.*

*Additional*

| | |
|---|---|
| $a_c$ | Crack size just tolerable at fatigue limit |
| $a_f$ | Final crack length |
| $a_o$ | Initial crack length |
| $C$ | Constant in crack growth equation |
| $K_c$ | Critical value of $K_I$ at which brittle fracture takes place |
| $l$ | Length over which $da/dN$ is averaged |
| $m$ | Exponent in crack growth equation |
| $N$ | Number of cycles |
| $r_p$ | Plastic zone radius |
| $\Delta K$ | Range of $K_I$ in fatigue cycle |
| $\Delta K_c$ | Threshold value of $\Delta K$ for fatigue crack growth |
| $\Delta K_o$ | Initial value of $\Delta K$ |
| $\Delta \sigma$ | Range of stress in fatigue cycle |

## 8.1 INTRODUCTION

The phenomenon of metal fatigue has been studied for over 100 years and now has a voluminous literature. Despite this, fatigue continues to be a major problem and is the most common cause of failures in engineering structures. The traditional approach to design against fatigue is to base allowable fatigue stresses on the results of tests on carefully-made plain or notched laboratory specimens, or on representative structures. The results of such tests are presented as $S/N$ curves, which relate cyclic stress to number of cycle to failure. Most materials exhibit a fatigue limit below which the life is indefinitely long. Numerous factors such as mean stress, environment, and surface finish affect a material's fatigue properties.

The main features of metal fatigue are now well understood **(8/1)**: a fatigue crack initiates at the surface and then grows until the net section is reduced and the part can no longer sustain its working load. A test on a plain or mildly notched specimen is largely a test of a material's resistance to crack initiation. Although it was pointed out in 1927 **(8/2)** that a full understanding of fatigue required a knowledge of crack growth behaviour virtually no experimental work was carried out before 1950. However it is now realized that most structures either contain crack-like defects introduced during manufacture, especially if welding is used, or develop them early, so that virtually the whole fatigue life is occupied by fatigue crack growth, and design needs to be based on the fatigue crack growth properties of the material concerned. The use of a traditional approach based on $S/N$ curves leads to difficulties because a conventional fatigue test does not give any information on the relative contribution of crack initiation and crack growth to total fatigue life. This can lead to difficulties in the understanding of the behaviour of sharply notched (where cracks initiate early) or cracked structures, in particular 'size effect' remains mysterious. The exceptions, for which a traditional approach based on tests on plain or mildly notched specimens is more appropriate, are usually carefully-made, highly stressed components.

As a result, testing to determine fatigue crack growth data is now widespread. The concept of stress intensity factor has proved to be a particularly convenient applied mechanics framework for the description and analysis of fatigue crack growth behaviour and for the solution of practictical engineering problems involving fatigue crack growth; consequently its use is now virtually universal. Because of mathematical difficulties, only limited attempts have been made to apply non-linear fracture mechanics to fatigue crack growth. This is not a serious limitation, principally because the size of the reversed plastic zone which appears on unloading is only one quarter of that for static loading. Only a limited number of references are cited; most of the topics mentioned are discussed in more detail either in ref. **(8/1)** or elsewhere in this special issue, and ref. **(8/3)** gives an historical survey.

A fairly rigorous statement on the current state of the applications of fracture mechanics to fatigue crack growth is that the concept of stress intensity factor can usefully be applied to the study of the macroscopic features of macrocrack growth (Stage II in Forsyth's terminology **(8/4)**); in this paper the term crack is taken to mean macrocrack. In principle fracture mechanics can also provide a framework for the study of microcrack (Stage I) growth, which takes place on planes of maximum shear stress. So far its use appears to be mainly descriptive; for example microcracks grow in Mode II or Mode III or both (Fig. 35).

On a macroscopic scale, fatigue fracture surfaces are generally flat and smooth in appearance. They tend to grow in Mode I (Fig. 35) irrespective of their initial orientation, so attention is largely confined to this mode. Other modes can occur when a crack follows a plane of weakness **(8/5)**, or in composite materials **(8/6)**. Crack growth in thin sheets is usually on 45° through the thickness planes (Fig. 36). These cracks are sometimes

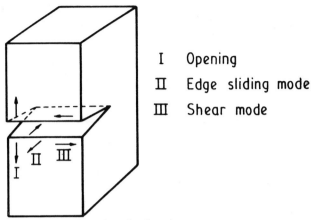

Fig. 35  Modes of crack surface displacement.

erroneously called shear cracks, because they are on planes of maximum shear stress in uncracked sheets, but they are actually a combination of Modes I and III. The direction of growth of a Mode I crack is approximately (exactly for symmetrical loadings) perpendicular to the maximum

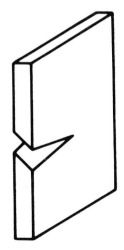

Fig. 36  Crack on 45° plane.

principal applied stress. A crack which has grown entirely in Mode I is not necessarily straight, and crack trajectories are not readily determined (8/7). As a general rule, a crack tends to be attracted by the nearest free surface and may follow a curved path even under initially symmetrical loading conditions. Cracks in structures are frequently found to follow complex paths. Ensuring that a crack maintains its initial direction is an important factor in the design of fracture mechanics test specimens.

Although more usually discussed for metals, the stress intensity factor concept is equally applicable to fatigue crack growth in thermoplastics (8/8). There are some differences in detail; the most significant is that more note must be taken of loading frequency. In composite materials complex crack patterns may develop leading to formidable mathematical problems, so that facture mechanics concepts cannot readily be applied (8/6).

## 8.2 DETERMINATION OF FATIGUE CRACK GROWTH DATA

Fatigue crack growth tests are relatively straightforward and over the past 20 years large numbers have been carried out. As already mentioned it is now the usual practice to analyse data in terms of stress intensity factors. These provide a particularly convenient means of correlating fatigue crack growth data, because the stress conditions at the crack tip can be described by a single parameter (8/9). In general the opening mode stress intensity factor is given by

$$K_\mathrm{I} = \sigma\, a^{\frac{1}{2}}\, Y \tag{1}$$

where $\sigma$ is applied stress, $a$ crack length and $Y$ stress intensity factor coefficient which corrects for geometric effects. The fatigue cycle is usually described by $\Delta K$, which equals $K_{\max} - K_{\min}$ where $K_{\max}$ and $K_{\min}$ are values of the opening mode stress intensity factor $K_\mathrm{I}$, calculated from the maximum and minimum stress during the fatigue cycle; if the minimum stress is compressive, it is conventionally taken as zero. This is a simplification based on the assumption that a crack closes when the load falls to zero. In practice a crack may close at above (8/10) or below (8/9) zero load. It has been shown experimentally that $\Delta K$ has the major influence on fatigue crack growth and, in general, if $\Delta K$ is constant, the fatigue crack growth rate is constant. For many materials the rate of fatigue crack growth can be expressed by the equation

$$\frac{da}{dN} = C(\Delta K)^m \tag{2}$$

where $N$ is the number of cycles, $C$ a material constant, and $m$ an exponent, usually about 3.

A plastic zone develops at the tip of a crack in a ductile metal, its approximate radius, $r_p$, being given by (**8/9**)

$$r_p = \frac{1}{2\pi} \left(\frac{K_\mathrm{I}}{\sigma_Y}\right) \tag{3}$$

for plane stress and one third of this amount for plane strain. $\sigma_Y$ is the yield stress (usually taken as the 0·2 per cent proof stress). Provided that $r_p$ is small compared with the crack length and that the maximum net section stress does not exceed 0·8 $\sigma_Y$ the plastic zone has little effect on the overall elastic stress field; if required, a correction can be made by adding $r_p$ to a when calculating $K_\mathrm{I}$. As the reversed yielding which takes place on unloading is only of limited extent a correction to $\Delta K$ is not usually necessary. If required it can be made using equation (3) but with $K_\mathrm{I}$ replaced by $\Delta K$ and $\sigma_Y$ by $2\sigma_Y$.

Equation (2) is sometimes modified (**8/11**) to allow for the increase in crack growth rate, which usually occurs as $K_{\max}$ approaches $K_c$ the critical value of $K_\mathrm{I}$ at which brittle fracture takes place, and for the effect of mean stress, which influences fatigue crack growth rates in some materials. The availability of a master curve relating $da/dN$ and $\Delta K$ enables a designer to predict growth rates for any cracked body configuration, and he is not limited to situations similar to those pertaining to the cracked specimen geometry used to generate the original data.

Numerous apparently different 'laws' of fatigue crack growth, including a number based on stress intensity factors, have been described in the literature (**8/11**): all the laws can be regarded as valid in the sense that they describe a particular set of fatigue crack growth data, and can consequently be used to predict crack growth rates in situations similar to those used to collect the data. It is sometimes possible to fit the same set of data to apparently contradictory laws but, owing to the inherent scatter in fatigue crack growth data, it is not possible to decide which law is the most 'correct' (**8/9**).

Table 2 summarizes some fatigue crack growth data (**8/12**) for a wide range of materials. In general, the fatigue crack growth properties of

Table 2  *Fatigue crack growth data for various materials.*

| Material | Tensile strength (MN/m²) | 0·1 or 0·2 per cent proof stress (MN/m²) | R | m | $\Delta K$ for $da/dN = 10^{-6}$ mm/c (MN/m$^{\frac{3}{2}}$) |
|---|---|---|---|---|---|
| Mild steel | 325 | 230 | 0·06–0·74 | 3·3 | 6·2 |
| Mild steel in brine* | 435 | — | 0·64 | 3·3 | 6·2 |
| Cold rolled mild steel | 695 | 655 | 0·07–0·43 | 4·2 | 7·2 |
| | | | 0·54–0·76 | 5·5 | 6·4 |
| | | | 0·75–0·92 | 6·4 | 5·2 |
| Low alloy steel* | 680 | | 0   –0·75 | 3·3 | 5·1 |
| Maraging steel* | 2010 | | 0·67 | 3·0 | 3·5 |
| 18/8 Austenitic steel | 665 | 195–255 | 0·33–0·43 | 3·1 | 6·3 |
| Aluminium | 125–155 | 95–125 | 0·14–0·87 | 2·9 | 2·9 |
| 5% Mg-Aluminium alloy | 310 | 180 | 0·20–0·69 | 2·7 | 1·6 |
| HS30W Aluminium alloy (1% Mg. 1% Si, 0·7% Mn) | 265 | 180 | 0·20–0·71 | 2·6 | 1·9 |
| HS30WP Aluminium alloy (1% Mg, 1% Si, 0·7% Mn) | 310 | 245–280 | 0·25–0·43 | 3·9 | 2·6 |
| | | | 0·50–0·78 | 4·1 | 2·15 |
| L71 Aluminium alloy (4·5% Cu) | 480 | 415 ⎫ | 0·14–0·46 | 3·7 | 2·4 |
| L73 Aluminium alloy (4·5% Cu) | 435 | 370 ⎭ | 0·50–0·88 | 4·4 | 2·1 |
| DTD 687A Aluminium alloy (5·5% Zn) | 540 | 495 | 0·20–0·45 | 3·7 | 1·75 |
| | | | 0·50–0·78 | 4·2 | 1·8 |
| | | | 0·82–0·94 | 4·8 | 1·45 |
| ZW1 Magnesium alloy (0·5% Zr) | 250 | 165 ⎫ | 0 | 3·35 | 0·94 |
| AM503 Magnesium alloy (1·5% Mn) | 200 | 107 ⎭ | 0·5 | 3·35 | 0·69 |
| | | | 0·67 | 3·35 | 0·65 |
| | | | 0·78 | 3·35 | 0·57 |
| Copper | 215–310 | 26–513 | 0·07–0·82 | 3·9 | 4·3 |
| Phosphor bronze* | 370 | | 0·33–0·74 | 3·9 | 4·3 |
| 60/40 brass* | 325 | | 0   –0·33 | 4·0 | 6·3 |
| | | | 0·51–0·72 | 3·9 | 4·3 |
| Titanium | 555 | 440 | 0·08–0·94 | 4·4 | 3·1 |
| 5% Al Titanium alloy | 835 | 735 | 0·17–0·86 | 3·8 | 3·4 |
| 15% Mo Titanium alloy | 1160 | 995 | 0·28–0·71 | 3·5 | 3·0 |
| | | | 0·81–0·94 | 4·4 | 2·75 |
| Nickel* | 430 | | 0   –0·71 | 4·0 | 8·8 |
| Monel* | 525 | | 0   –0·67 | 4·0 | 6·2 |
| Inconel* | 650 | | 0   –0·71 | 4·0 | 8·2 |

\* Data of limited accuracy obtained by an indirect method.

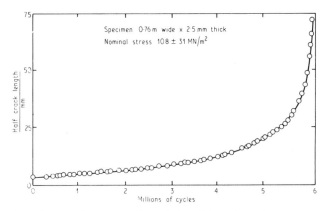

*Fig. 37 Growth curve for a central crack in mild steel specimen.*

steels are broadly independent of type, yield stress ($\sigma_Y$), or the value of $R$, where $R$ is the ratio of the minimum to maximum stress in a fatigue cycle. Fatigue crack growth data obtained by different investigators on similar materials usually agree reasonably well although there may be considerable variation in the amount of scatter (**8/13**). Obviously some scatter arises from experimental inaccuracies. However although fatigue fractures appear smooth, on a microscopic scale fatigue crack growth is a very irregular process, and this also contributes to scatter in data.

Fatigue crack growth rates can be determined from a wide range of specimens, including all the types used for fracture toughness testing (**8/12**). Tests are usually carried out under constant amplitude loading, with measurements of crack length and number of cycles made at intervals. Fig. 37 shows some actual test results (**8/1**) for a mild steel centre-cracked sheet. Such basic data are analysed by calculating values of $\Delta K$ and $da/dN$ for various crack lengths. Data from one or more specimens are then used to determine the relationship between $\Delta K$ and $da/dN$. The values of $da/dN$ are obtained either by direct calculation between successive pairs of readings or from the slope of a curve fitted to the basic $a$ versus $N$ data. Scatter often makes the first method impractical and the second difficult. However

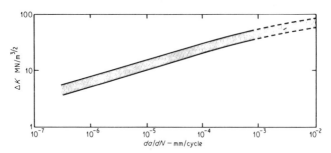

*Fig. 38  Crack growth data for mild steel.*

the matter is approached, subjective judgements are involved; for example use of computerized curve fitting techniques involves judgement in selecting the basic approach and details of the procedure. Collaborative test programmes have been carried out in Britain and USA (**8/14**) with a view to developing a recommended practice, but a completely satisfactory procedure has yet to be evolved (**8/13**). A standardized procedure of course has the advantage of eliminating personal error.

Numerous different methods can be used to monitor crack length during a fatique test. The commonest methods are optical observation of the specimen surfaces and the potential drop technique. In the potential drop technique, a high constant current is applied through leads at the specimen ends; potential leads are located at each side of the crack to measure the change in potential drop across the crack as it grows. There is usually considerably more scatter in data obtained on thin sheets using optical crack length measurement than in data obtained on thick specimens using the potential drop technique (**8/13**). Figure 38 shows, plotted on logarithmic scales, fatigue crack growth data (**8/9**) obtained by optical measurement on thin mild steel sheets; the scatter band was drawn to include 90 per cent of the data. A simple statistical calculation (**8/13**) suggests that the width of the scatterband is proportional to $1/l^{\frac{1}{2}}$ where $l$ is the effective distance over which an average crack growth rate is determined. This effective distance is increased by data smoothing techniques. Further the scatterband is wider for optical measurement on the surfaces than when a technique such as the

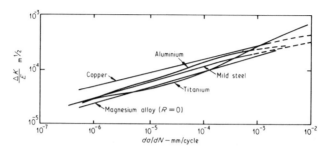

*Fig. 39  Crack growth data for various materials compared on basis of strain.*

potential drop method, which can be arranged to measure the average crack length over the whole crack front, is used.

Most of the band in Fig. 38 is straight so that its centre line can be represented by equation (1); the dotted portion at the top indicates data where the maximum stress on the uncracked portion is in the region of general yield. Provided that the stress range is small compared with the mean stress, $\Delta K$ can still be used to correlate data in this region; the rationale is that unloading deformations will be largely elastic. For some materials net section yielding causes an increase in crack growth rate. When tests are carried out it is important (8/9) that data obtained when the maximum net section stress is greater than $0\cdot 8\, \sigma_Y$ are identified. Fatigue crack growth is basically a strain controlled process so that materials of different Young's moduli show similar behaviour when they are compared on the basis of strain instead of stress, as shown in Fig. 39 where crack growth rates for various materials are plotted against $\Delta K/E$.

Equation (1) predicts that any value of $\Delta K$, no matter how small, will cause crack growth. However, if $\Delta K$ is below a certain threshold level, $\Delta K_c$, fatigue crack growth does not take place (8/9): thus, for a crack of given length, a minimum fatigue stress is necessary for crack growth. This threshold is associated with a fatigue crack growth rate of around one lattice spacing/cyle (about $4 \times 10^{-7}$ mm/c), which is the minimum possible on physical grounds. Lower average rates are sometimes observed, particularly in corrosive environments; in such cases, crack growth can be taking place only on part of the crack front during each cycle. Unlike the rate of crack growth, which is often independent of $R$, $\Delta K_c$ normally decreases as $R$ increases. For many materials $K_c$ is approximately proportional to $\{(1-R)/(1+R)\}^{\frac{1}{4}}$. $\Delta K_c$ values for various materials (**8/12, 8/15**) are shown in Tables 3 and 4. For a given $R$ value $\Delta K_c$ is largely independent of a steel's strength or composition. Immersion in brine substantially reduces $\Delta K_c$ for mild steel and somewhat reduces $\Delta K_c$ for aluminium alloy.

Thresholds can be obtained by a variety of techniques. The most obvious

*Table 3   Values of $\Delta K_c$ for various ferrous materials.*

| Material | Tensile strength (MN/m²) | $R$ | $\Delta K_c$ (MN/m$^{\frac{3}{2}}$) |
|---|---|---|---|
| Mild steel | 430 | —1 | 6·4 |
|  |  | 0·13 | 6·6 |
|  |  | 0·35 | 5·2 |
|  |  | 0·49 | 4·3 |
|  |  | 0·64 | 3·2 |
|  |  | 0·75 | 3·8 |
| Mild steel at 300°C | 480 | —1 | 7·1 |
|  |  | 0·23 | 6·0 |
|  |  | 0·33 | 5·8 |
| Mild steel in brine | 430 | —1 | ~2·0 |
|  |  | 0·64 | 1·15 |
| Mild steel in brine with cathodic protection | 430 | 0·64 | 3·9 |
| Mild steel in tap water or SAE30 oil | 430 | —1 | 7·3 |
| Low alloy steel | 835 | —1 | 6·3 |
|  | 680 | 0 | 6·6 |
|  |  | 0·33 | 5·1 |
|  |  | 0·50 | 4·4 |
|  |  | 0·64 | 3·3 |
|  |  | 0·75 | 2·5 |
| NiCrMoV steel at 300°C | 560 | —1 | 7·1 |
|  |  | 0·23 | 5·0 |
|  |  | 0·33 | 5·4 |
|  |  | 0·64 | 4·9 |
| Maraging steel | 2010 | 0·67 | 2·7 |
| 18/8 Austenitic steel | 685 | —1 | 6·0 |
|  | 665 | 0 | 6·0 |
|  |  | 0·33 | 5·9 |
|  |  | 0·62 | 4·6 |
|  |  | 0·74 | 4·1 |
| Grey cast iron | 255 | 0 | 7·0 |
|  |  | 0·50 | 4·5 |

Table 4  Values of $\Delta K_c$ for various non-ferrous materials.

| Material | Tensile strength (MN/m²) | $R$ | $\Delta K_c$ (MN/m$^{\frac{3}{2}}$) |
|---|---|---|---|
| Aluminium | 77 | —1 | 1·0 |
|  |  | 0 | 1·7 |
|  |  | 0·33 | 1·4 |
|  |  | 0·53 | 1·2 |
| L65 Aluminium alloy | 450 | —1 | 2·1 |
| (4·5% Cu) | 495 | 0 | 2·1 |
|  |  | 0·33 | 1·7 |
|  |  | 0·50 | 1·5 |
|  |  | 0·67 | 1·2 |
| L65 Aluminium alloy |  | 0·50 | 1.15 |
| (4·5% Cu) in brine* |  |  |  |
| ZW1 Magnesium alloy | 250 | 0 | 0·83 |
| (0·6% Zr) |  | 0·67 | 0·66 |
| AM503 Magnesium | 165 | 0 | 0·99 |
| alloy (1·6% Mn) |  | 0·67 | 0·77 |
| Copper | 225 | —1 | 2·7 |
|  | 215 | 0 | 2·5 |
|  |  | 0·33 | 1·8 |
|  |  | 0·56 | 1·5 |
|  |  | 0·69 | 1·4 |
|  |  | 0·80 | 1·3 |
| Phosphor bronze | 325 | —1 | 3·7 |
|  | 370 | 0·33 | 4·1 |
|  |  | 0·50 | 3·2 |
|  |  | 0·74 | 2·4 |
| 60/40 brass | 330 | —1 | 3·1 |
|  | 325 | 0 | 3·5 |
|  |  | 0·33 | 3·1 |
|  |  | 0·51 | 2·6 |
|  |  | 0·72 | 2·6 |
| Titanium | 540 | 0·60 | 2·2 |
| Nickel | 455 | —1 | 5·9 |
|  | 430 | 0 | 7·9 |
|  |  | 0·33 | 6·5 |
|  |  | 0·57 | 5·2 |
|  |  | 0·71 | 3·6 |
| Monel | 525 | —1 | 5·6 |
|  |  | 0 | 7·0 |
|  |  | 0·33 | 6·5 |
|  |  | 0·50 | 5·2 |
|  |  | 0·67 | 3·6 |
| Inconel | 655 | —1 | 6·4 |
|  | 650 | 0 | 7·1 |
|  |  | 0·57 | 4·7 |
|  |  | 0·71 | 4·0 |

* Unpublished data

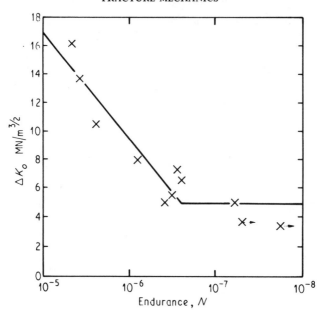

*Fig. 40   Tests on cracked NiCrMoV steel plates at 300°C.*

is simply to follow the $da/dN$ against $\Delta K$ curve downwards. However, unless this is done very carefully the threshold can be seriously overestimated and some published data obtained this way are suspect as, in general, are data obtained by extrapolating a fitted curve to zero crack growth rate. A straightforward, if somewhat tedious, technique is to determine an $S/N$ curve for cracked specimens, with lives plotted against the initial value of $\Delta K$, $\Delta K_o$, rather than stress. Figure 40 shows some results (**8/15**) for NiCrMoV steel tested at 300°C.

A feature of fatigue crack growth in thin sheets is the transition to growth on a 45° plane (Fig. 36) which often takes place (**8/16**) as crack length increases. The reason for this transition is not completely clear, but it appears to be associated with a change from plane strain to plane stress conditions; it has been correlated with the attainment of a critical value of either $\Delta K$ or $da/dN$ (**8/1**). Evidence is conflicting on whether the transition causes an increase or decrease in crack growth rate, but in any case the difference is not great. Indeed thickness in general has little effect on fatigue crack behaviour, although for critical applications it is advisable to use data for material of about the thickness used in the structure.

The effect of environment on fatigue crack growth rates is complex and individual problems must be treated on their merits (8/1). In general corrosive environments increase the rate of crack growth particularly at low loading rates. Examination of published data (8/17) suggests that the influence of temperature on fatigue crack growth rates is largely through its influence on Young's modulus and yield stress. Sometimes metallurgical factors influence crack growth rates by causing a change in fracture mechanism.

## 8.3 APPLICATION OF FATIGUE CRACK GROWTH DATA TO CONSTANT AMPLITUDE PROBLEMS

In a structure containing crack-like flaws virtually the whole of the fatigue life is occupied by fatigue crack growth characteristics of the material from which it is made. For steel these are independent of the steel's strength. The fatigue strength of the structure is therefore also largely independent of the strength of the steel from which it is made. This explains why the use of higher strength steels with concomitant higher working stresses results in an increased likelihood of fatigue failure.

A common problem is to determine the number of cycles needed for a crack in a component to grow from some initial length $a_o$ to the final length $a_f$ at which either the conditions for a brittle fracture are satisfied or the section is so reduced that the working stress exceeds the material's tensile strength. Substituting equation (1) into equation (2) and integrating gives the number of cycles needed to grow a crack from $a_o$ to $a_f$,

$$N = \frac{1/\{a_o^{(m/2)-1}\} - 1/\{a_f^{(m/2)+1}\}}{C\Delta\sigma^m \{(m/2 - 1)\} Y^m} \quad (4)$$

where $\Delta\sigma$ is the range of stress during the fatigue cycle but neglecting any compressive stresses. Usually $a_f$ will be large compared with $a_o$ and equation (4) can be simplified to

$$N = \frac{1}{a_o^{(m/2)-1} C\Delta\sigma^m \{(m/2) - 1\} Y^m} \quad (5)$$

Equation (5) provides a simple explanation of 'size effect' for any structure containing a crack-like flaw. For example if the flaw size remains in proportion the effect of increasing size is to decrease the number of cycles to failure at a given stress level.

Equations (4) and (5) assume that $Y$ remains constant as a crack grows, but when accurate calculations are necessary allowance must be made for the increase in $Y$ with crack length which usually takes place; the integration then usually has to be carried out numerically. The actual expression used to represent fatigue crack growth data does not greatly affect the predicted number of cycles provided that it fits the data adequately and is not extrapolated beyond the range of the original data. Errors near the final crack length where the crack is growing fast are less important than in the early stages of crack growth where most of the cycles to failure are accumulated, as can be seen from the form of equation (4). The expected, or most probable number of cycles will be obtained if the mean line through the crack growth data is used for prediction. Scatter about this life will be small as, in effect, an average rate is being taken over a large number of cycles (**8/13**). The mean line should therefore be used in applications such as failure analysis where the expected life is required.

In practice scatter bands for fatigue crack growth data often incorporate data obtained from different batches of nominally similar materials, and part of the scatter can be attributed to batch-to-batch variations in material behaviour. It is not in general possible to separate the scatter due to material variation from that due to the random nature of fatigue crack growth (**8/13**). It is often implicitly assumed that all the scatter is due to variations in material behaviour. This conservative assumption is commonly employed in safety assessments by using the upper bound of the scatter band in calculations. Variations in fatigue crack growth rates appear directly as variations in calculated lives, and difficulties sometimes arise (**8/18**) because of a wide discrepancy between the expected life and that which can be assured when allowance is made for the various uncertainties, including uncertainties in load history and stress intensity factors for postulated or detected flaws.

Diagrams relating $N$, $\Delta\sigma$, and $a_0$ can readily be prepared for various design situations, although experienced judgement may be needed in making the necessary simplifications. Figure 41 shows an example (**8/12**), based on equation (4), where the effect of forging laps on the fatigue life of a component at its working load has been estimated for three positions: appropriate values of $\Delta K_c$ have also been used to estimate the non-propagating flaw sizes. Diagrams of this nature are being increasingly used

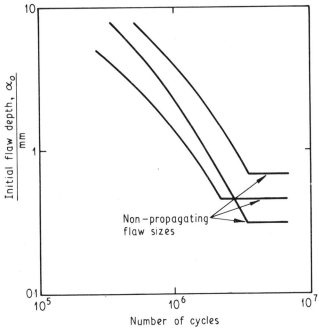

*Fig. 41 Effect of forging laps in three positions on the fatigue life of a component at normal working load.*

for applications such as estimating the maximum flaw size which can be tolerated in a structure. Various scaling factors derived from equations (4) and (5) have proved particularly useful in the understanding of the fatigue behaviour of welded joints (**8/19**). An important point that follows from the existence of the threshold for fatigue crack growth, and is illustrated by Figure 41, is that, in the absence of a corrosive environment, a cracked component normally has a definite fatigue limit, with a knee in the $S/N$ curve at around $10^6$–$10^7$ cycles. It follows from equation (4) that the slope of the $S/N$ curve for a cracked structure plotted on logarithmic scales is $-1/m$, and this relationship is being taken into account in the revision of design stresses for welded joints (**8/20**).

The determination of the threshold stress necessary for crack growth is particularly important for components such as engine parts, which are subjected to very large numbers of cycles during service: once a crack starts to grow, the component will inevitably fail. Consider, for example, the case of a large diesel engine crank-shaft which failed after several hundred hours running. Examination showed that failure was caused by a fatigue crack which originated at a forging lap about 2 mm deep in a filler. Calculations (**8/12**) showed that $\Delta K$ for the flaw was above $\Delta K_c$, so it was clear that the forging lap, rather than any abnormal operating conditions, caused the failure. However in a case involving fatigue cracks which had grown from badly finished radii in marine steam turbine blades, it was found that stresses of an order of magnitude greater than the design stresses were needed to cause crack growth. It was concluded that cracking was due to a combination of bad workmanship and a previously unsuspected resonance.

The existence of a threshold for fatigue crack growth implies that a material can tolerate cracks of up to a certain critical size before the plain fatigue limit is affected. Values of the critical crack size $a_c$ which various materials can just tolerate at their fatigue limits are typically (**8/12**) in the range of 0·005 to 0·25 mm; for mild steel it is 0·06 mm. If a material contains inherent crack-like flaws, such as casting defects, which are above the critical size, these must be taken into account when fatigue data obtained from plain specimens are interpreted. For example Fig. 42 shows test results (**8/12**) for cast bronze of a type used for ship's propellers. The results were obtained from four-point bending tests at a mean stress of 124 MN/m². When plotted conventionally (Fig. 42a) there was considerable scatter, with an ill-defined fatigue limit at about $124 \pm 60$ MN/m² ($R \approx \frac{1}{3}$). Examination of the fracture surfaces showed that failure had originated at irregular crack-like flaws varying in size from 0·32 to 3·05 mm. The initial value of $\Delta K$, $\Delta K_o$, was estimated for each specimen, and the results replotted (Fig. 42b). The scatter was now much reduced, and there was a reasonably well defined fatigue limit at a $\Delta K_o$ of about 4 MN/m$^{\frac{3}{2}}$, corresponding to $\Delta K_c$ for the material. This was in reasonable agreement with $\Delta K_c$ for other copper alloys with $R \approx \frac{1}{3}$ (Table 4). It is therefore clear that the flaws were controlling the fatigue behaviour of the material and that failure took place at the most severe flaw (the one with the highest $\Delta K$) in a particular specimen.

## 8.4 THE VARIABLE AMPLITUDE PROBLEM

Most structures in service are subjected to varying amplitude loads and this must be taken into account in predictions. Variable amplitude loadings can

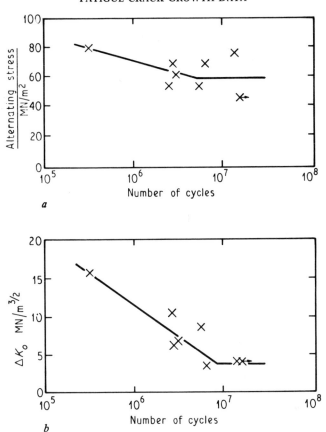

*Fig. 42  S/N curves for 25·4 mm square cast bronze bars tested in four point bending.*

be divided into two broad classes—those in which individual load cycles may be distinguished, for example narrow band random loading (**8/19**), and those in which individual cycles cannot be distinguished, for example broad band random loading (**8/21**).

The order in which loads of varying amplitude are applied can have a profound influence on rates of crack growth (**8/16, 8/22, 8/23**). In general a fatigue crack grows at the expected rate when the load is increased but is usually retarded for a time following load reductions. Basically, inter-

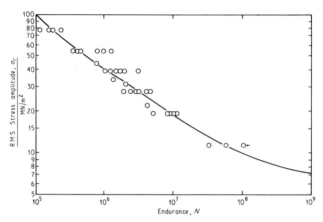

*Fig. 43 Test results for mild steel load carrying partial penetration fillet welds tested under narrow band random loading at mean stress of 154 MN/m² compared with prediction based on constant amplitude data.*

actions occur because of the compressive residual stresses which arise at a crack tip when a load is removed, the wake of plastically deformed material adjacent to the crack surfaces which may cause crack closure before the minimum load is reached, and sometimes because of the incompatible crack front orientations which arise when component cycles are above and below the transition to 45° crack growth (**8/16**). At present the effect of interactions between different load levels cannot be predicted theoretically so an empirical model in which the constants are adjusted to suit the results of tests must be used (**8/22**).

In narrow band random loading, a cycle does not differ greatly from its predecessor; this reduces interaction effects, which are relatively unimportant in low strength steels. Thus for low strength steels under narrow band random loadings it is possible to make the simple assumption that each cycle causes the same amount of growth as if it were applied as part of a sequence of loads of constant amplitude (**8/19, 8/23**). More generally however, this simple approach can result in excessively conservative predictions (**8/17**). In calculations it is convenient to calculate the amount of

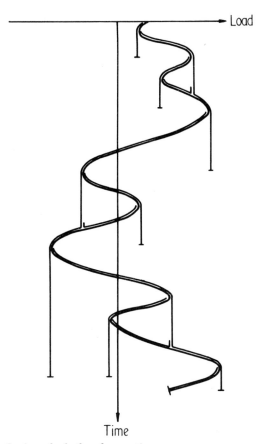

*Fig. 44* '*Rainflow*' *method of cycle counting.*

crack growth caused by the variable amplitude loading with that caused by a constant amplitude loading and having the same root mean square (rms) stress amplitude and mean stress calculations and then simulate those for a constant amplitude problem (**8/19**). As a variation, constant amplitude fatigue data for the structure of interest can be used to estimate variable amplitude lives by fracture mechanics methods. Figure 43 illustrates the good agreement which can be obtained for welded joints (**8/19**).

The situation is further complicated when individual cycles cannot be distinguished. Various methods of calculating equivalent cyclic spectra have been developed (**8/21**). but all have shortcomings; the most logically defensible and sophisticated is the 'rainflow' method, in which rain is imagined to flow down a load time record plotted vertically (Fig. 44). Flow starts at the beginning of the record, then the inside of each peak in the order in which peaks are applied. It stops when it either meets flow from a higher level, or a point opposite a peak which is arithmetically greater (or equal to) the point from which it started, or when the end of the record is reached. Each separate flow is counted as a half cycle. There is always a complementary half cycle of opposite sign, except perhaps for a flow which either starts at the beginning of the record or reaches the end.

The advent of servo-hydraulic fatigue testing equipment, with which virtually any desired load history can be applied to a structure or test specimen (**8/24, 8/25**), has considerably simplified the determination of life under variable amplitude loading. Methods of analysis, such as the rainflow method, in conjunction with appropriate fracture mechanics when structures contain crack-like flaws, then become models which permit experimental data under one set of conditions to be extrapolated with confidence to broadly similar situations, the optimum approach to a particular problem being a judicious blend of theoretical calculations, experiment, and analysis of relevant service experience. One useful approach is the use of realistic load histories to determine fatigue crack growth data with the data obtained then used to predict the behaviour of structures. Results could conveniently be analysed in terms of the rms values of $K_I$. This would automatically take interaction effects into account, without the need for tests on full scale structures.

## 8.5   CONCLUSIONS

1. In fatigue problems a distinction must be made between situations where crack-like flaws are absent, which can readily be dealt with by a traditional approach based on $S/N$ curves, and situations where crack-like flaws are present so that virtually the whole fatigue life is occupied by crack growth from a flaw, for which a fracture mechanics approach is appopriate. The latter include most welded joints.

2. A large amount of fatigue crack growth data has been accumulated, and the fracture mechanics concept of stress intensity factor provides a particularly convenient method of analysing such data so that they are in a form which can be applied directly to engineering problems.

3. Accumulated data and analysis have led to a much better understanding of the fatigue behaviour of structures containing crack-like flaws, especially welded joints. The fracture mechanics approach is now used extensively at the design stage and in failure analysis.

4. A major unsolved problem in fatigue crack growth is a full understanding of the effects of interaction between different load levels. This is not necessarily a serious drawback for practical engineering purposes as servo-hydraulic fatigue testing equipment permits the application of virtually any desired load history to representative structures or fatigue crack growth specimens.

## 8.6  ACKNOWLEDGEMENT

This paper is published by permission of the Director, National Engineering Laboratory, Department of Industry, and is Crown copyright.

# 9
# Environmental Effects in Crack Growth
## R. N. PARKINS
*Department of Metallurgy and Engineering,*
*The University of Newcastle-upon-Tyne*

The interdependence of material structure, electrochemical parameters and the response of a material to the application of stress can produce slow crack growth by a number of different mechanisms. The recognition of the importance of such factors as environment composition, especially within the confines of a pit or crack, and of plasticity, especially in relation to the rate at which deformation presents bare metal at the tips of cracks, is vital if reproducible data are to be obtained.

### 9.1 INTRODUCTION

Interactions between materials and environments can result in slow crack growth at stress intensities markedly below those associated with fast fracture, the fracture surface characteristics frequently, but not invariably, showing marked differences from those observed in the absence of any environmental interaction. Thus, Fig. 45 shows the fracture surface associated with ductile failure in a 12 per cent Cr steel, which is typical of such failure in other metals, together with examples of intergranular fracture in a 12 per cent Cr steel produced by contact with tap water and of transgranular fracture in a Mg–7 per cent Al alloy produced by contact with an aqueous chromate-chloride solution. Such fractographic characteristics, together with the knowledge that crack branching is likely to occur in the event of environmental interaction, frequently allow the diagnosis of failure by stress corrosion cracking. Environments that promote slow crack growth are frequently highly specific to the material concerned, in that not all environments promote cracking. This requirement of specific solutions

*The MS of this paper was received at the Institution of Mechanical Engineers on 16 May 1975 and accepted for publication on 22 August 1975.*

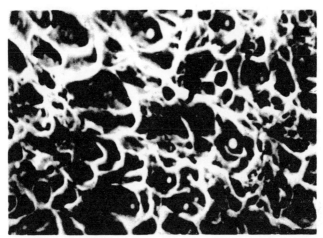

(a) Scanning electron micrograph of fracture surface of 12 per cent Cr steel fractured in air.

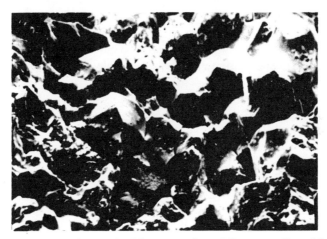

(b) Scanning electron micrograph of fracture surface of 12 per cent Cr steel fractured in tap water.

coupled with the compositional and structural parameters of a susceptible material vary so widely from system to system that rationalization of all of these factors in a single explanation would be difficult if not unreal, i.e. it is probable that a number of different mechanisms are involved (**9/1**).

(c) Scanning electron micrograph of fracture surface of Mg–7 per cent Al alloy fractured in a chromate–chloride solution.

*Fig. 45*

The usual energy balance approach to fracture needs modification where corrosion processes are involved to take account of the chemical energy released (9/2) which distinguishes environmental sensitive fracture from other modes of failure. Thus,

| Surface energy change | + | Plastic work done near crack tip | = | Change in initial stored energy | + | Electrochemical energy released | |
|---|---|---|---|---|---|---|---|
| | | | | | | | (1) |

The surface energy component usually will be negligible in comparsion with the plastic work component ($\gamma_p$) in the stress corrosion of ductile materials and, hence

$$\gamma_p = K_I^2(1 - v^2)/E + (zF\rho\delta/M)\eta \tag{2}$$

where $z$ is the valency of the solvated ions, $F$ is Faraday's constant, $\rho$ is the density, $\delta$ is the height of the advancing crack front (approximating to the COD), $M$ is the molecular mass of the metal and $\eta$ is the anodic overpotential. At the threshold stress intensity, $K_{Iscc}$, i.e. the minimum value of $K_I$ for stress corrosion cracking, equation (2) yields

$$K_{Iscc} = \{(E/1 - v^2)(\gamma_p - zF\rho\delta\eta/M)_{min}\}^{\frac{1}{2}} \tag{3}$$

Clearly the variables that may influence $K_{Iscc}$, and hence the susceptibility to environmental sensitive fracture, are $\gamma_p$ and $\eta$

i.e. $K_{Iscc} = \{k_1(\gamma_p - k_2\eta)_{min}\}^{\frac{1}{2}}$ (4)

A reduction of the plastic work term, $\gamma_p$, will result from an increase in the effective yield stress or from an increase in the work hardening rate in the crack tip region, either of which, for constant $\eta$, will lower $K_{Iscc}$ and hence increase the susceptibility to stress corrosion, as also will an increase in the anodic overpotential. $\eta$ will be some function of the electrochemical conditions within the crack, i.e. of pH, anion activity, metal composition and electrode potential. The interdependence of these terms upon the structure and composition of the metal, upon the details of the electrochemical conditions at the crack tip in terms of local cell action and film formation and upon the response of the metal to the presence of stress in creating new metal at the crack tip, make the analysis of the detailed mechanisms of crack propagation virtually impossible, as West (9/2) indicates. However, the recognition of the need for a critical balance between a number of variables if stress corrosion is to occur, and the fact that this balance may be achieved in a number of ways, is important, and not least in relation to the diagnosis and the prevention of environmentally sensitive fracture.

## 9.2 STRESS CORROSION CRACK PROPAGATION MODELS

The implication of the foregoing, that stress corrosion cracking will occur if a mechanism exists for concentrating the electrochemical energy release rate at the crack tip or if the environment in some way serves to embrittle the metal provides a convenient introduction to a consideration of the mechanistic models of environmentally sensitive crack growth. If the geometrical requirements of a crack are to be fulfilled in the presence of a corrosive environment, metal dissolution must be localized at the crack tip and the rate of anodic dissolution may be expressed as a rate of crack propagation

$$V = \frac{i_a M}{zF\rho}$$ (5)

where $i_a$ is the anodic current density. Now $i_a$ will be dependent upon the nature of the phase being dissolved, upon the balancing cathodic processes that occur elsewhere and upon the properties of the environment, but how does the imposition of a tensile stress influence the situation? For corrosion to proceed along a narrow front retention of the geometry of a crack implies that most of the exposed surfaces, including the crack sides, must remain relatively inactive. The transition from electrochemically active to relatively inactive behaviour that the sides of a crack must undergo as the tip advances and creates more crack can only be achieved if the environment reacts to form a film. This implies in relation to equation (5) that the conditions for stress corrosion will be met if $i_a$ is maintained close to the film-free value, i.e. if protective films are not allowed to grow over the crack tip or, if this does occur, that the film is repetitively broken. The function of stress thus will be essentially to prevent, or to fracture, films forming at the crack tip.

Two different circumstances may now be envisaged whereby cracks can propagate by a dissolution controlled process. The alloy may exhibit structural features, either as a segregate or as a precipitate, usually at the grain boundaries, that cause a local galvanic cell to be established. Where such pre-existing active paths are non-existent, or are inoperative, the disruption of a protective surface film to expose bare metal may result in a second mechanism of crack propagation, the active path along which the crack propagates being cyclically generated as disruptive strain and film build-up occur sequentially so that propagation is related to the slip characteristics of the underlying metal. The significance of alloy structure is inherent in these mechanisms, as also are the compositions of the metal and the environment and hence the concept of solution specificity. Thus, the combination of metal and environment will determine whether or not an electrochemically active $\rightarrow$ passive transition can occur whereby lateral dissolution at the crack sides is prevented whilst crack tip dissolution continues, and, if so, defining the range of electrode potentials at which cracking will occur. Reasonable agreement (9/3) is obtained between observed potential ranges for cracking and those predicted from appropriate electrochemical measurements, and also between observed crack velocities and those calculated from equation (5).

The above refers to processes under dissolution control, but the energy balance of equation (1) indicates that, with negligible contribution from dissolution, crack extension will be facilitated by a reduction in the surface energy required to form crack faces or by a reduction in the plastic work term by embrittlement of the metal in the crack tip region. If the environment provides species that are adsorbed at the crack tip to reduce the effective bond strength, then the surface energy is effectively lowered. Alternatively, the species may diffuse into the metal forming a brittle phase, e.g. a hydride, at the crack tip, or adsorption may occur at some legion in advance of the crack tip where the stress and/or strain conditions are particularly appropriate for the nucleation of a crack. In the latter case hydrogen is usually regarded as the only species that can diffuse with sufficient speed to account for observed crack propagation rates, the hydrogen being derived from a corrosion reaction or simply from a gaseous atmosphere. Surface energy lowering has been suggested (9/4) as a single mechanism that explains all instances of environmentally sensitive fracture, including those involving ceramic and polymeric materials, for which there is some supporting evidence (9/5) although with the latter materials plasticization by organic agents in producing crazing appears to be the dominant process (9/6). However, surface energy lowering cannot easily explain environmentally sensitive fracture in the more ductile materials. Thus, whilst stress corrosion cracks can propagate without marked macroscopic plastic deformation there is ample evidence to show that localized plastic deformation occurs at the crack tip and in such circumstances, as indicated by equation (1), the surface energy term is negligible in relation to the plastic work term (5 $J/m^2$ as opposed to 5 $kJ/m^2$) and so any reduction of surface energy by adsorption will have a negligible effect upon the fracture stress.

Where the mechanism of cracking involves embrittlement of the metal in the crack tip region a strain energy argument is involved and this implies in relation to equation (1) that plastic strain should be minimized and elastic energy maximized for failure, conditions that are most readily met with high yield strength materials. It is well established that the hydrogen embrittlement of steels becomes more marked the higher the yield strength, although changes in structure or composition that result in a change in yield strength, or fracture toughness, may also influence electrochemical reactions and such parameters as hydrogen diffusivity and these may be as significant as any change in strength in influencing stress corrosion behaviour. The nature of the solutions that affects cracking in materials of low ductility is not particularly specific, e.g. high strength steels will fail in a wide range of aqueous and organic solutions, unlike the situation in relation to the failure of the low strength, ductile alloys. The common factor in these environments is hydrogen and the requirement is simply that the hydrogen can pass into the metal, so that species in solution which facilitate the ingress of hydrogen into the metal will enhance cracking, whilst species that lead to the discharge of gaseous hydrogen at the steel surface before it can enter the metal will retard cracking. In the former category are sulphur and arsenious salts, whilst platinum additions to the system may be expected to facilitate hydrogen discharge. Similarly the effect of increasing cathodic current densities applied in stress corrosion tests may be expected to enhance cracking if hydrogen adsorption is involved in the failure mechanisms and all of these anticipated effects are observed (**9/7**). The demonstration that sub-atmospheric pressures of hydrogen gas can readily result in the propagation of cracks in high strength steels indicates that the mechanism is not likely to involve the diffusion of hydrogen through the metal to voids where a disruptive pressure of gas is generated. Of the alternatives that (i) hydrogen lowers the surface energy by adosrption or (ii) that it accumulates within a few atomic distances from the crack tip in response to the lowering of its chemical potential by the elastic component of stress, thereby lowering the cohesive force of the lattice, Oriani (**9/8**) prefers the latter because it is the only mechanism that is consistent with the observations of the effects upon crack propagation of small changes in hydrogen gas pressure and of the substitution of deuterium for hydrogen. A sufficient reduction in the hydrogen gas pressure surrounding a WOL specimen, containing a propagating crack at a given stress intensity, caused the crack to stop propagating, but a subsequent increase in pressure, of about 12 torr from 165 torr, was sufficient to restart the crack and with a delay time that was so short that the extra hydrogen entering the lattice could have diffused no more than a few atom spacings. Similarly rapid responses of the crack velocity to small changes in cathodic current applied to an 18 per cent Ni maraging steel immersed in NaCl have been observed. The effect of deuterium in reducing the response to embrittlement appears to be unrelated to the difference in the transport kinetics of the two isotopes but to their solubilities in the dilated lattice just beyond the crack tip. This again is in agreement with the decohesion model, in contrast to one involving surface energy lowering, and is further supported by the observation (**9/9**) that a stress field having a dilatant component, produced by mode I tensile loading, leads to hydrogen induced cracking in a high strength steel, whereas at even higher stress intensities, mode III loading (anti-plane shear mode), which has a zero dilatant component, does not respond to hydrogen induced cracking.

The suggestion (**9/1**) that these different mechanisms of stress corrosion occur within a continuous spectrum, with a gradual transition from one to the other as the dominance of corrosive processes is replaced by stress or strain, leads readily to the notion that alloy composition and structure, electrochemistry and stress may interact in a variety of ways and that the transformation from one mechanism to another may result from a change in either alloy characteristics or environmental conditions.

## 9.3 THE EFFECTS OF STRESS INTENSITY

Figure 46 shows (**9/10**) the relationship between stress intensity factor and stress corrosion crack velocity for a high strength steel in distilled water and is typical of the trends that have been observed in other systems (**9/5, 9/6, 9/11, 9/12**) and for a variety of loading conditions, net section stresses, specimen shapes and crack lengths. The curves are characterized by a strongly dependent region of the crack velocity (Region I), followed by a plateau (Region II) in which the crack velocity is independent of stress intensity and a Region III in which the crack velocity is again stress dependent. More complicated curves are sometimes observed, showing a gradual transition from Regions I to II or two plateaux at crack velocities differing by orders of magnitude. Quantifiable explanations of Regions I, II and III are still awaited, although it appears likely that Region II is under electrochemical control and Region III represents a transition to fast mechanical fracture leading to total failure at $K_{Ic}$. Region I effectively relates to the threshold stress intensity for stress corrosion cracking, $K_{Iscc}$ and defines the critical crack size below which significant crack growth does not occur, i.e. for a surface crack in an infinite plate remotely loaded in tension (**9/13**)

$$a_{cr} = 0.2 \left(\frac{K_{Iscc}}{\sigma_Y}\right)^2 \tag{6}$$

where $a_{cr}$ is the shallowest crack, long at the surface compared to its depth, that will just propagate as a stress corrosion crack. In those systems where, for crack extension, anodic processes need to occur at a film free crack tip, Region I and $K_{Iscc}$ may be expected to relate to the conditions that provide a film free tip, i.e. to plastic deformation, at a rate that will sustain an appropriate balance between the filming rate and the rate at which bare metal is created. There are various indications suggesting the importance of crack tip strain rate.

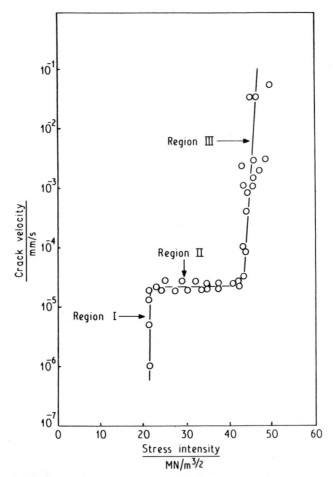

*Fig. 46 Effect of applied stress intensity upon crack velocity for high strength ($180 \times 10^9$ N/m² UTS) quenched and tempered steel (AFC 77) in distilled water (9/10).*

## 9.4 STRAIN RATE EFFECTS

Stress corrosion tests **(9/14)** upon a Mg–7 per cent Al alloy in a chromate–chloride solution have shown a simple relationship between the threshold stress for the failure of initially plain specimens and $K_{Iscc}$ for pre-cracked

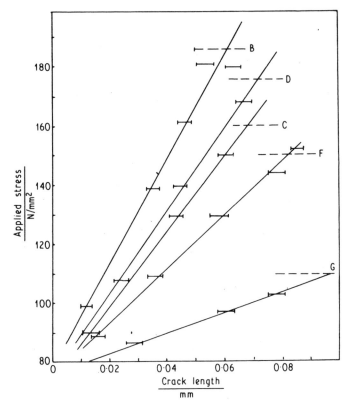

*Fig. 47 Maximum non-propagating crack depths for different initial stresses applied to a Mg–7 per cent Al alloy whilst immersed in a chromate–chloride solution. The threshold stresses are shown as broken lines for the various structural conditions produced by different heat treatments (9/13).*

specimens. Insofar as the latter defines a critical crack size below which propagation to total failure does not occur, the relationship with the threshold stress for initially plain specimens prompts the question as to whether for these also the critical flaw size concept is applicable. Since the plain specimens did not initially contain flaws of appropriate size, the implication is that stress corrosion cracks initiate and propagate to some extent before arresting if the stress is below the threshold value, whereas above the latter propagation continues to total failure. Such non-propagating cracks are indeed observed below the threshold stress in initially plain specimens, their length being dependent upon the initial stress and the structural condition of the alloy in the manner shown in Fig. 47. Moreover, the crack length and the threshold stress are quadratically related, indicating that the threshold stress determined upon plain specimens can be explained by fracture mechanics. However, more intriguing is the question why cracks, having initiated, stop propagating, notwithstanding the increasing stress intensity as the crack depth increases. The answer appears to be concerned with the effective strain rate, since the threshold stress is sensitive to the relative times at which the stress is applied and at which the environmental conditions for cracking are established.

Similar results to those described above have been obtained with C–Mn steels immersed in a carbonate–bicarbonate solution, non-propagating cracks being observed below the threshold stress for initially plain specimens and cracking being sensitive to the effective strain rate at the time the electrochemical conditions for cracking are established. Figure 48 shows the cracking responses of precracked specimens as reflected in the deflection of the beam whilst subjected to constant load as cantilevers. (The specimen dimensions did not conform to the requirements for linear elastic analysis to be valid, but the COD determined from the beam deflection was within about 25 per cent of the COD calculated from the measured crack extension.) Upon loading such a specimen creep occurs in the plastic zone but if the conditions are such that no stress corrosion crack propagation takes place, e.g. by controlling the electrode potential at a non-cracking value, creep exhaustion will eventually occur and the beam displacement with time follows a simple creep law (Curve A in Fig. 48). However, if the electrochemical conditions promote stress corrosion cracking (Curve B) the beam deflection rate eventually accelerates because the crack extension at constant load leads to an increased stress intensity and hence to more creep. The deflection of the beam therefore affords a sensitive method of detecting crack growth, that is later confirmed metallographically. Curve C shows the result of changing the potential when the creep rate has fallen to a relatively low value and the subsequent beam deflection indicates that stress corrosion cracking did not occur, since Curve C is not essentially different from Curve A. Curve D shows the effect of changing the potential at a much earlier stage, when the creep rate is still relatively high, and the subsequent beam deflection behaviour indicates crack growth, confirmed metallographically. Since all of these tests were performed at the same initial net

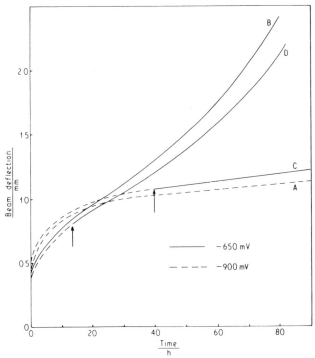

*Fig. 48 Effects of time of change of potential upon stress corrosion crack propagation reflected in beam deflection behaviour of precracked cantilever loaded specimens of C–Mn steel immersed in a carbonate–bicarbonate solution. (−650 mV (sat. calomel electrode) is a potential that promotes cracking but at −900 mV cracking does not occur) (9/15).*

section stress the different cracking responses are more readily interpreted in terms of the effect of creep rate than of any direct influence of stress.

A more convincing demonstration of the importance of strain rate is obtained from tests in which the strain rate is superimposed, rather than allowed to vary as is inevitable in constant load tests. Experiments upon the same system as that to which Fig. 48 refers, but with the beam displaced at various rates by a device that replaces the conventional load pan, produced the type of results shown in Fig. 49. (The specimens were pre-loaded and allowed to creep at a non-cracking potential until creep was effectively exhausted, when the beam deflection rate was applied in conditions where

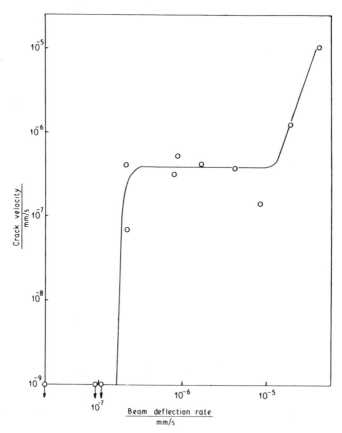

*Fig. 49 Effect of applied beam deflection rate upon stress corrosion crack velocity in C-Mn steel immersed in carbonate–bicarbonate solution at a cracking potential (9/15).*

the changes in effective stress during the tests were negligible, except at the highest deflection rate.) The results clearly indicate a lower limiting strain rate below which crack propagation is not observed, followed by a region in which the intergranular stress corrosion crack velocity is independent of strain rate and then, at relatively high strain rates, a transition to fast transgranular tearing. The similarity to the results shown in Fig. 46 is noteworthy and especially so in view of the observation (**9/15**) that stress corrosion cracking in another high strength steel is sensitive to the loading rate.

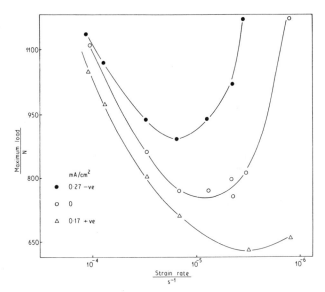

*Fig. 50 Effects of various applied strain rates upon the stress corrosion cracking behaviour of a Mg–7 per cent Al alloy immersed in a chromate–chloride solution with various applied current densities. (The maximum load achieved is a measure of the cracking propensity, ductile failures occurring at loads in excess of 1100 N).*

The range of strain rates within which stress corrosion cracking is observed in the system to which Fig. 49 refers is dependent upon the electrochemical conditions, as also is the plateau crack velocity (9/16). Similar effects are observed in the Mg–Al alloy referred to earlier in that, in tensile tests upon initially plain specimens, stress corrosion cracking is observed only if the strain rate lies within a clearly defined range, outside of which, at higher or lower strain rates, ductile fracture ensues, the boundaries of the critical range being dependent upon the electrochemical conditions in the manner shown in Fig. 50. The maintenance of film-free conditions at the crack tip are therefore not only a function of the strain rate but also of the film growth rate as determined by the electrochemical conditions. The frequent observation of a relationship between yield stress and threshold stress (**9/16**) for plain specimens in lower strength steels and the association of stretch zones with the initiation of environmental sensitive cracking from pre-cracks (**9/17**) in higher strength steels are further results that support a model involving these complementary functions of stress and electrochemistry.

## 9.5 ELECTROCHEMICAL EFFECTS IN GEOMETRICAL DISCONTINUITIES

The frequent observation that stress corrosion cracks initiate from corrosion pits has often been taken to indicate that pitting constitutes the initiation stage for environmentally sensitive fracture, because of the local stress configuration that is generated by the growth of a pit, and hence that the use of pre-cracked specimens circumvents the initiation stage. However, it appears at least as likely that geometrical discontinuities, whether in the form of pre-cracks or corrosion pits, will be as important for electrochemical reasons as for those concerned with their influence upon stress distribution. Thus, within a discontinuity, environmental conditions may develop which, because mixing is effectively prevented, may be markedly different from those represented by the bulk of the environment and indeed in some cases it is probable that the conditions for cracking do not exist within the bulk conditions but are self generated within a discontinuity. There is a considerable amount of evidence (**9/18**) which shows that changes in electrode potential, pH and solution composition occur within pits, and Smith *et al* (**9/19**) have made similar measurements in relation to the stress corrosion cracking of a high strength steel. In the latter work the pH and the electrode potential were measured at the tips of propagating cracks for various bulk solution pH values and conditions of external polarization. Figure 51 shows the data obtained, superimposed upon a potential-pH diagram for the $Fe-H_2O$ system. Clearly the crack tip pH is virtually independent of the bulk pH and is determined solely by the electrochemical reactions occurring within the confines of the crack. Moreover, the crack tip potential, irrespective of the conditions of external polarization, invariably falls below the line that defines the limit of thermodynamic stability of water, i.e. the crack tips were at potentials where water would decompose to provide a source of hydrogen which entered the metal to promote fracture. Control of the pH or potential so that the crack tip conditions

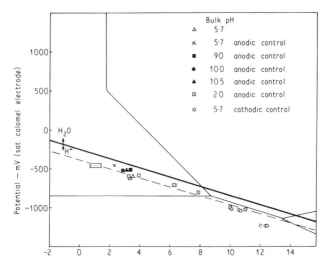

*Fig. 51* Crack tip potential and solution pH values observed in the stress corrosion of a high strength steel in chloride solutions of varying bulk pH and with various conditions of external electrochemical control (9/18).

remained above the line relating to the decomposition of water resulted in no crack growth. Clearly, apart from any other considerations, such results indicate the importance of geometrical discontinuities in providing circumstances in which electrochemical conditions for cracking may be generated.

## 9.6 CRACK MORPHOLOGY

Stress corrosion cracks rarely adopt simple straight-sided, sharp-tipped, morphologies but frequently show evidence of blunting or branching that complicate their analysis. Blunting or the termination of a crack in a pit of diameter appreciably greater than the crack tip width are likely to occur if there is a change in the electrochemical conditions within the crack, but branching may be a more predictable phenomenon in that it appears to occur at specific stress intensities (**9/20**). Microbranching, in which short cracks deviate from the major crack but cease to propagate as the major crack continues, occurs at a minimum stress intensity of about 1·4 $K_{Iscc}$. Macrobranching, involving the bifurcation of an initially single crack, only occurs in Region II of the type of curve shown in Fig. 46 and, designating the minimum stress intensity for the constant crack velocity plateau as $K_p$, macrobranching occurs when the stress intensity exceeds 1·4 $K_p$. If $K_p$ is very low, as with some of the very ductile materials, single cracks are rarely observed and multiple branching is common unless the initial stress intensity is extremely low. However, despite the complications that might be expected to be associated with stress corrosion crack branching, it is conceivable that measurements of the stress intensity at branching could provide a value for $K_{Iscc}$ without recourse to multiple specimen testing, although this requires a knowledge of the factor relating $K_{Iscc}$ to the stress intensity for branching but this, currently, probably introduces too large an error for many purposes.

# 10
# Fracture Mechanics: A Summary of its Aims and Methods

## 10.1 INTRODUCTION

The object of this closing paper is not so much to review the 'state of the art', but to summarize the viewpoint, methodology and scope of fracture mechanics so that a coherent picture can be seen of what fracture mechanics sets out to achieve and how it achieves it, unclouded by the niceties and uncertainties of assumptions and approximation inherent in the foregoing more detailed descriptions of particular aspects of the subject.

## 10.2 THE FRACTURE MECHANICS VIEWPOINT

Fracture mechanics supposes the pre-existence of a significant crack-like defect that will lead to failure. Elastic or elastic–plastic continuum mechanics theory is used to develop parameters which characterize the crack tip stress field; the most important of these is the stress intensity factor, $K$. If a material property is measured in terms of a suitable parameter under known circumstances, fracture mechanics then allows this property to be used to calculate the behaviour of other geometrical configurations for similar physical circumstances. If the property is concerned with brittle fracture, then it is 'fracture toughness'; it could however, be concerned with fatigue crack growth, stress corrosion, or even creep cracking. The material may be metallic, polymeric, ceramic or a composite since the microscopic nature of the fracture process is not discussed.

## 10.3 THE USAGE OF FRACTURE MECHANICS

Fracture mechanics has led to the emergence of new design concepts, and efforts are being made to incorporate these in standards and codes of practice. In particular small cracks may be accepted provided the component is still fit for its design purpose; indeed the attempted repair of a defect can sometimes be more harmful than its acceptance. The introduction of the concept of the usage of a flawed component has no counterpart in conventional design. The only way the relationship between stress level and resistance to crack growth can be used is by specifying a flaw size, and this by its very nature forms no part of normal design processes. In other words continuity between the design method for flawed or unflawed components is lacking.

In practice, two levels of usage tend to emerge. The first covering perhaps 90 per cent of practical problems in a simple 'go' or 'no-go' prediction. The crudest assessment of $K$ and rough knowledge of material properties such as $K_{Ic}$ may well be adequate to indicate the acceptability or otherwise of a particular postulated or detected flaw.

At the other extreme very careful examination of local stress fields (including residual and thermal stresses), three dimensional effects, and adjacent defects may be necessary to estimate $K$. Equally material properties such as $K_{Ic}$ may be affected by degradation through welding, some form of metallurgical embrittlement, anisotropy and material variability. The high technology industries have tended to establish very thorough procedures for quality control, inspection and non-destructuve testing as a necessary workshop complement to the usage of design methods based on fracture mechanics.

If a small defect is accepted, then knowledge of its growth rate, generally by fatigue, will allow the logical definition of a re-inspection cycle time adequate to detect the larger crack before it has grown enough to cause fracture. Similarly, if there is uncertainty about the size of crack that may have escaped detection, then survival of an overload test demonstrates that cracks above a certain size cannot exist, or they would have caused fracture. Hence at operating load a margin exists between possible and acceptable crack size. Fracture mechanics can assist in the selection of the optimum material or heat treatment for a particular job. For example the right balance between strength and toughness can be specified, and reasonable absolute values established for a particular purpose, thus avoiding either costly early failure or the use of unnecessarily expensive material.

In high technology problems, where perhaps an actual defect has been detected, knowledge is at present often found to be barely adequate to make a final assessment of the margins on which acceptance or rejection of a very expensive piece of hardware may be based. Reference (**10/1**) gives a good general account of the amount of effort which can be involved in the solution of an apparently straightforward problem.

## 10.4 CONCLUSIONS

The strength of fracture mechanics is the separation of geometrical features from mechanical properties in the presence of cracks, so that unique values are defined for given circumstances, and these can then be used in assessing the behaviour of components. It offers a logical basis for extrapolations into hitherto unused sizes of components or strength ranges of materials. A state of development useful in design applications has been reached, but that is not to say that future developments will not occur to supplement or replace existing ideas, since much yet remains to be understood about the problem of fracture.

# General Reference Section

(1/1)   GRIFFITH, A. A. 'The phenomena of rupture and flow in solids', *Phil. Trans. Roy. Soc.* 1921 **A221**, 163.

(1/2)   BUECKNER, H. F. 'Propagation of cracks and the energy of elastic deformation', *Trans. Am. Soc. Mech. Engrs.* 1958 **80E**, 1225.

(1/3)   MUSHKELISHVILI, N. I. *Some basic problems of the mathematical theory of elasticity* Translated by J. R. M. Radok 1963 (P. Noordhoff Ltd, Groningen).

(1/4)   GRIFFITH, A. A. 'The theory of rupture', *Proc. 1st Int. Congr. Appl. Mech.* Delft, 1924.

(1/5)   IRWIN, G. R. 'Fracture dynamics', in *Fracturing of metals* ASM Cleveland, 1948.

(1/6)   OROWAN, E. 'Fracture and strength of solids', *Rep. Prog. Phys.* 1949 **12**, 185.

(1/7)   OROWAN, E. 'Energy criteria of fracture', *Weld. J. Res. Suppl.* 1955 **20**, 157s.

(1/8)   FELBECK, D. K. and OROWAN, E. 'Experiments on brittle fracture of steel plates', *Weld. J. Res. Suppl.* 1955 **34**, 570s.

(1/9)   IRWIN, G. R. and KIES, J. A. 'Fracturing and fracture dynamics', *Weld. J. Res. Suppl.* 1952 **17**, 95s.

(1/10)  LOVE, A. E. *A treatise on the mathematical theory of elasticity* 4th ed. 1944 (Dover, New York).

(1/11)  DAY, E. E. 'Strain energy release rate determination for some perforated structural members', *Weld. J. Res. Suppl.* 1956 **21**, 60s.

(1/12)  SRAWLEY, J. E., JONES, M. H. and GROSS, B. 'Experimental determination of the dependence of crack extension force on crack length for a single-edge-notch tension specimen', 1964 NASA TND-2396.

(1/13)  IRWIN, G. R. 'Analysis of stresses and strains near the end of a crack traversing a plate', *J. Appl. Mech.* 1957 **24**, 361.

(1/14)  IRWIN, G. R. 'Relation of stresses near a crack to the crack extension force', *9th Int. Congr. Appl. Mech.* Brussels, 1957.

(2/1)   Ibid. (1/13).

(2/2)   Ibid. (1/14).

(2/3)   WESTERGAARD, H. M. 'Bearing pressures and cracks', *J. Appl. Mech.* 1939 **61**, A49.

(2/4)   SIH, G. C. 'Application of strain-energy-density theory to fundamental fracture problems', *Technical Report AFOSR-TR-73-1*, 1973.

(2/5)   WIGGLESWORTH, L. A. 'Stress distribution in a notched plate', *Mathematika* 1957 **4**, 76.

## GENERAL REFERENCE SECTION

(2/6) WILLIAMS, M. L. 'On the stress distribution at the base of a stationary crack', *J. Appl. Mech.* 1957 **24**, 109.

(2/7) ERDOGAN, F. 'On the stress distribution in plates with colinear cuts under arbitrary loads', *Proc. 4th U.S. Natn. Congr. on Appl. Mech.* Berkeley, 1962.

(2/8) IRWIN, G. R. 'Crack extension forces for a part through crack in a plate', *J. Appl. Mech.* 1962 **29**, 651.

(2/9) PARIS, P. C., SIH, G. C. and ERDOGAN, F. 'Crack tip stress intensity factors for plane extension and plate bending problems', *J. Appl. Mech.* 1962 **29**, 306.

(2/10) SIH, G. C. 'On the singular character of thermal stresses near a crack tip', *J. Appl. Mech.* 1962 **29**, 587.

(2/11) BOWIE, O. L. 'Rectangular tensile sheet with symmetric edge cracks', *J. Appl. Mech.* 1964 **31**, 208.

(2/12) PARIS, P. C. and SIH, G. C. 'Stress analysis of cracks' in *Fracture toughness testing and its applications* 1965 (ASTM STP 381, Philadelphia).

(2/13) SIH, G. C. 'Stress distribution near internal crack tips for longitudinal shear problems', *J. Appl. Mech.* 1965 **32**, 51.

(2/14) KNOWLES, J. K. and WANG, N. 'On the bending of an elastic plate containing a crack', *J. Math. and Phys.* 1960 **39**, 223.

(2/15) SIH, G. C. and HAGENDORF, H. C. 'A new theory of spherical shells with cracks' in *Thin-shell structures* Ed. by Fung, Y. C. and Sechler, E. E. 1974 (Prentice-Hall, Englewood Cliffs).

(2/16) BERGOZ, D. and RADENKOVIC, D. 'On the definition of stress intensity factors in cracked plates and shells', *2nd Int. Conf. on Pressure Vessel Technology*, San Antonio, 1973.

(2/17) IRWIN, G. R. 'Fracture mode transition for a crack traversing a plate', *J. Basic Engng.* 1960 **82**, 417.

(2/18) LIU, H. W. Discussion in *Fracture toughness testing and its applications* 1965 (ASTM STP 381, Philadelphia).

(2/19) LARSSON, S. G. and CARLSSON, A. J. 'Influence of non-singular stress terms on small scale yielding at cracks tips in elastic-plastic materials', *3rd Int. Conf. on Fracture*, Munich, 1973.

(2/20) TADA, H. *The stress analysis of cracks handbook* 1973 (Del. Research Corp. Hellertown).

(2/21) BROWN, W. F. and SRAWLEY, J. E. *Plane strain crack toughness testing* 1967 (ASTM STP 410, Philadelphia).

(3/1) *Ibid.* (2/1) 163–198.

(3/2) KNOTT, J. F. *Fundamentals of fracture mechanics* 1st Edition 1973 (Butterworths, London).

(3/3) British Standards Institution, Draft for Development DD3. 'Methods for plane strain fracture toughness ($K_{Ic}$) testing'.

(3/4) WILSON, W. K. Contribution to discussion 'Plane strain crack toughness testing of high strength metallic materials' ASTM STP 410, 1969, 75–76.
(3/5) DUGDALE, D. S. 'Yielding of steel sheets containing slits', *Jnl. Mech. Phys. Solids* 1960 **8**, 100–104.
(3/6) BILBY, B. A., COTTRELL, A. H. and SWINDEN, K. H. 'The spread of yield from a notch', *Proc. Roy. Soc.* 1963 **A272**, 304–314.
(3/7) BURDEKIN, F. M. and STONE, D. E. W. 'The crack opening displacement approach to fracture mechanics in yielding materials', *Jnl. Strain Analysis* 1966 **1**, 145–153.
(3/8) RICE, J. R. 'Crack tip plasticity and fracture initiation criteria' Third International Congress on Fracture, Munich 1973, paper 441.
(3/9) *Ibid.* (2/19).
(3/10) RICE, J. R. 'Limitations to the small-scale yielding approximation for crack tip plasticity', *Jnl. Mech. Phys. Solids*, 1974 **22**, 17–26.
(3/11) LEWIS, I. D., SMITH, R. F. and KNOTT, J. F. 'On the $a/W$ ratio in plane strain fracture toughness testing', *Int. Journ. of Fracture* 1975 **11**, 179–183.
(3/12) RICE, J. R. 'A path independent integral and the approximate analysis of strain concentration by notches and cracks', *Jnl. Appl. Mechanics* 1968 (June) 379–386.
(3/13) HEALD, P. T., SPINK, G. M. and WORTHINGTON, P. J. 'Post yield fracture mechanics', *Mat. Sci. Eng.* 1972 **10**, 129–138.
(3/14) CHELL, G. C. and KIRBY, J. to be published, Metals Technology.
(3/15) BEGLEY, J. A. and LANDES, J. D. 'The effect of specimen geometry on $J_{Ic}$', ASTM STP 514, 1972, 24–39.
(3/16) KRAFFT, J. M., SULLIVAN, A. M. and BOYLE, R. W. 'Effects of dimensions on fast fracture instability of notched sheets', Crack Propagation Symposium, Cranfield 1961, Paper 1.
(3/17) KNOTT, J. F. 'Some effects of hydrostatic tension on the fracture behaviour of mild steel', *Jnl. Iron and Steel Inst.* 1966 **204**, 104–111.
(3/18) TRACEY, D. M., Ph.D. Thesis, Brown University 1973.
(3/19) RITCHIE, R. O., RICE, J. R. and KNOTT, J. F. 'On the relationship between critical tensile stress and fracture toughness in mild steel', *Jnl. Mech. Phys. Solids* 1973 **21**, 395–410.
(3/20) CURRY, D. A. and KNOTT, J. F. 'The relationship between microstructure and low temperature fracture toughness in mild steel' submitted to *Metal Science*.
(3/21) GREEN, G. and KNOTT, J. F. 'The effect of specimen size on the ductile/brittle transition temperature in A533B pressure vessel steel', presented at the conference, 'Materials defects of steel products', Karlovy Vary, Czechoslovakia 17–19 Sept. 1975.

(3/22) GENIETS, L. C. E., RITCHIE, R. O. and KNOTT, J. F. 'Effects of grain-boundary embrittlement on fracture and fatigue crack propagation in a low-alloy steel'. 'The microstructure and design of alloys' Third Intl. Conf. on Strength of Metals and Alloys, Cambridge 1973, *Inst. Metals*/I.S.I., 124–128.

(3/23) GREEN, G. and KNOTT, J. F. 'The initiation and propagation of ductile fracture in low strength steels', presented at conference 'Micromechanical modelling of deformation and fracture', ASME Tray 23–25 June 1975 and to be published (JEMT).

(3/24) ELLIOTT, D. 'Crack tip processes leading to fracture' in 'The practical implications of fracture mechanisms', Inst. of Metallurgists 1973, 21–27.

(3/25) GREEN, G., SMITH, R. F. and KNOTT, J. F. 'Metallurgical factors in low temperature slow crack growth', Conference on Mechanics and Mechanisms of Crack Growth, Cambridge 1973, British Steel Corporation, paper 5.

(3/26) RICE, J. R. and JOHNSON, M. A. 'The role of large crack tip geometry changes in plane strain fracture', 'Inelastic behaviour of solids' ed. M. F. Kannirren et al., McGraw-Hill 1970, 641–661.

(3/27) CHIPPERFIELD, C. G. and KNOTT, J. F. 'Microstructure and toughness of structural steels', *Metals Technology* 1975 **2**, 45–51.

(3/28) CLAYTON, J. Q. and KNOTT, J. F. 'Observation of fibrous fracture modes in a prestrained low alloy steel', Submitted to Metal Science.

(3/29) HAHN, G. T. and ROSENFIELD, A. R. 'Sources of fracture toughness: the relationship between $K_{IC}$ and the ordinary tensile properties of metals', Applications Related Phenomena in Titanium Alloys ASTM STP 432, 1968, 5–32.

(3/30) GARRETT, G. G., Ph.D. Thesis, Cambridge reported in KNOTT, J. F., Proc. Conference 'Mechanics and physics of fracture', IOP/Metals Society, Cambridge 1975, paper 9.

(3/31) BURDEKIN, F. M. and DAWES, M. G. 'Practical use of linear elastic and yielding fracture mechanics with particular reference to pressure vessels', *Proc. Inst. Mech. Engrs. Conf.* on 'Practical applications of fracture mechanics to pressure vessel technology', 1971, 28–37.

(4/1) HAYES, D. J. This work, chapter 2.

(4/2) McCLINTOCK, F. *Proc. Roy. Soc. (London)* Ser. A., 1965 **285**, 58.

(4/3) HILL, R. *The Mathematical Theory of Plasticity*. Oxford Univ. Press, 1950.

(4/4) *Ibid.* (3/5).

(4/5) BARENBLATT, G. I. 'The mathematical theory of equilibrium cracks in brittle fracture', *Advances in Applied Mechanics* 1962 **7**, 55.

(4/6) WELLS, A. A. 'Unstable crack propagation in metals: cleavage and fast fracture'. Crack Propagation Symposium, Cranfield, 1961.

(4/7) BURDELIN, F. M. and STONE, D. E. W. *J. Strain Anal.* 1966 **1**, (2), 145.

(4/8) BILBY, B. A., COTTRELL, A. H., SMITH, E. and SWINDEN, K. H. *Proc. Roy. Soc. A* 1964 **279**, 1.
(4/9) ROSENFIELD, A. R., DAI, P. K. and HAHN, G. T. *Proc. 1st Int. Conf. on Fracture*, Sendai, 1965 **1**, 223, Japanese Soc. for Strength and Fracture of Materials, Tokyo.
(4/10) WELLS, A. A. *J. Eng. Fract. Mech.* **1**, (3), (1968–70), 399.
(4/11) *Ibid.* (3/13).
(4/12) HAYES, D. J. and WILLIAMS, J. G. *Int. J. Fract. Mech.* 1972 **8**, 239.
(4/13) NICHOLS, R. W., BURDEKIN, F. M., COWAN, A., ELLIOTT, D. and INGHAM, T. *Practical Fracture Mechanics for Structural Steel* (Proc. Conf. Culcheth, 1969), UKAEA/Chapman and Hall, 1969, Section F.
(4/14) SMITH, R. F. and KNOTT, J. F. *Practical Application of Fracture Mechanics to Pressure Vessel Technology, Inst. Mech. Engrs.*, London, 1971, 65.
(4/15) KANAZAWA, T., MACHIDA, S. and MIYATA, T. *Prospects of Fracture Mechanics*, Ed. Sih, van Elst, Broek. Noordhoff Inst. (Leyden) 1974, 547.
(4/16) PRIEST, A. H. Present paper on toughness testing.
(4/17) *Ibid.* (3/31).
(4/18) DAWES, M. G. 'Fracture control in high yield strength weldments', *Weld. Res. Suppl.* Sept. 1974.
(4/19) Document X-WGSD-13, 1974. *Int. Inst. Weld. Assessment of the significance of defects.*
(4/20) HUTCHINSON, J. W. *J. Mech. Phys. Solids* 1968 **16**, 13.
(4/21) RICE, J. R. and ROSENGREN, G. F. *J. Mech. Phys. Solids* 1968 **16**, 1.
(4/22) *Ibid.* (3/12). 370.
(4/23) HAYES, D. J. This issue Paper No. 2.
(4/24) ESHELBY, J. D. *Solid State Physics* **3**, Academic Press, New York 1956, 79.
(4/25) CHERGPANOV, G. P. *J. App. Math. Mech.* 1967 **31**, 503.
(4/26) McCLINTOCK, F. *Fracture, An Advanced Treatise* **3**, (Ed. Liebowitz), Academic Press, New York, 1971, Chap. 2.
(4/27) LARSSON, S. G. and CARLSSON, A. F. *J. Mech. Phys. Solids* 1973 **21**, 263.
(4/28) HILTON, P. D. and SIH, G. C. *Mechanics of Fracture* **1**, *Methods of analysis and solutions of crack problems*. (Ed. Sih), Noordhoff, 1973, 426.
(4/29) MILLER, K. J. and KFOURI, A. P. *Int. J. Fract.* 1974 **10**, 393.
(4/30) HAYES, D. J. *Some applications of elastic-plastic analysis to fracture mechanics.* Ph.D. Thesis, University of London, 1970.
(4/31) BOYLE, E. F. *Calculations of elastic and plastic crack extension forces.* Ph.D. Thesis, Queen's University, Belfast, 1972.
(4/32) SUMPTER, J. D. G. *Elastic-plastic fracture analysis and design using the finite element method.* Ph.D. Thesis, London University, 1974.

| | |
|---|---|
| (4/33) | BUCCI, R. J., PARIS, P. C., LANDES, J. D. and RICE, J. R. STP 514, ASTM, Philadelphia, 1972, 40. |
| (4/34) | *Ibid.* (3/13). |
| (4/35) | LANDES, J. D. and BEGLEY, J. A. STP 514, ASTM, Philadelphia, 1972, 24. |
| (4/36) | RICE, J. R., PARIS, P. C. and MERKLE, J. G. *Progress in flaw growth and fracture toughness testing.* STP 536, ASTM, Philadelphia, 1973. |
| (4/37) | *Ibid.* (3/16). |
| (4/38) | ROBINSON, J. N. and TETELMAN, A. S. STP 559, ASTM, Philadelphia. |
| (4/39) | ROBINSON, J. N. 'An experimental investigation of the effect of specimen type on the crack tip opening displacement and $J$ integral fracture criterion'. (To be published.) |
| (4/40) | RICE, J. R. *Fracture, an advanced treatise*, Ed. Liebowitz, **2**, Academic Press, New York, 1971, Chapter 3. |
| (4/41) | BURDEKIN, F. M. and TURNER, C. E. *Atomic energy review* 1974 **12**, (3), 439. |
| (5/1) | *Ibid.* (1/1) 163–198. |
| (5/2) | *Ibid.* (1/4). |
| (5/3) | *Ibid.* (1/5) 147–166. |
| (5/4) | OROWAN, E. 'Fatigue and fracture of metals', MIT Symposium 1950, pp. 139–167, Wiley 1952. |
| (5/5) | *Ibid.* (2/12). |
| (5/6) | TADA, H., PARIS, P. and IRWIN, G. 'The stress analysis of cracks handbook', DEL Research Corporation, Hellertown, Pennsylvania 1973. |
| (5/7) | SIH, G. C. 'Handbook of stress intensity factors', Lehigh University, Bethlehem, Pennsylvania 1973. |
| (5/8) | ROOKE, D. P. and CARTWRIGHT, D. J. 'Compendium of stress intensity factors', HMSO 1975. |
| (5/9) | LIEBOWITZ, H. (Ed.) 'Fracture', 1–7, Academic Press 1968. |
| (5/10) | SIH, G. C. (Ed.) 'Methods of analysis and solution of crack problems', Noordhoff 1973. |
| (5/11) | CARTWRIGHT, D. J. and ROOKE, D. P. 'Methods of determining stress intensity factors', Technical Report TR73031, Royal Aircraft Establishment, Farnborough (1973). |
| (5/12) | *Ibid.* (2/3). |
| (5/13) | IRWIN, G. R. 'Fracture' *Handbuch der Physik VI*, 551–590, Springer-Verlag, Berlin (1958). |
| (5/14) | IRWIN, G. R. 'Structural mechanics', 557–594, Pergamon, New York (1960). |
| (5/15) | TADA, H. 'Westergaard stress functions for several periodic crack problems', *Engng fracture Mech.* 1970 **2**, 177–180. |
| (5/16) | *Ibid.* (1/3). |

(5/17) SIH, G. C. 'Application of Mushkelishvili's method to fracture mechanics', *Trans. Chinese Assoc. for Advanced Studies* 1962 **3**, 25.
(5/18) *Ibid.* (2/7).
(5/19) SAVIN, G. N. 'Stress distribution round holes', NASA Technical Translation, NASA TT F-607 1970.
(5/20) *Ibid.* (2/13).
(5/21) SHIRIAEV, E. A. 'Torsion of circular bars with two cuts', *Prikl. Mat. Mekh.* 1958 **22**, 549–553.
(5/22) *Ibid.* (2/6).
(5/23) TIMOSHENKO, S. D. and GOODIER, J. N. 'Theory of elasticity'. 3rd Ed. McGraw-Hill 1970.
(5/24) GROSS, B. and SRAWLEY, J. E. 'Stress intensity factors for single edge notch specimens in bending or combined bending and torsion by boundary collocation of a stress function', NASA TN D-2603 1965.
(5/25) GROSS, B., SRAWLEY, J. E. and BROWN, W. F. 'Stress intensity factors for a single edge notch tension specimen by boundary collocation of a stress function', NASA TN D-2395 1964.
(5/26) GROSS, B. and SRAWLEY, J. E. 'Stress intensity factors for three point bend specimens by boundary collocation', NASA TN D-3092 1965.
(5/27) GROSS, B. and SRAWLEY, J. E. 'Stress intensity factors by boundary collocation for single-edge-notch specimens subject to splitting forces', NASA TN D-3295 1966.
(5/28) GROSS, B. and SRAWLEY, J. E. 'Stress intensity factors for crackline-loaded edge-crack specimens', NASA TN D-3820 1967.
(5/29) WILLIAMS, M. L. 'The stresses around a fault or crack in dissimilar media', *Bull. Seismological Soc. of America* 1959 **49**, 199–204.
(5/30) SIH, G. C. and RICE, J. R. 'The bending of plates of dissimilar materials with cracks', *J. Appl. Mech.* 1964 **31**, 477–490.
(5/31) SAWYER, S. G. 'A stress intensity factor approach to the analysis of interfacial cracks in fibre reinforced composite material', Ph.D. Thesis, Carnegie-Mellon University.
(5/32) KOBAYASHI, A. S., CHEREPY, R. D. and KINSEL, W. C. 'A numerical procedure for estimating the stress intensity factor of a crack in a finite plate', *J. Bas. Engg.* 1964 **86**, 681–684.
(5/33) VOOREN, J. V. 'Remarks on an existing numerical method to estimate the stress intensity factor of a straight crack in a finite plate', *J. Bas. Engg.* 1967 **89**, 235–237.
(5/34) NEWMAN, J. C. 'An improved method of collocation for the stress analysis of cracked plates with various shaped boundaries', NASA TN D-6376 1971.
(5/35) ISIDA, M. and ITAGAKI, Y. 'Stress concentration at the tip of a central transverse crack in a stiffened plate subject to tension', *Proc. 4th U.S. National Congress of Appl. Mech.* 1962 **2**, 955–969.

(5/36) ISIDA, M. 'Stress intensity factors for the tension of an eccentrically cracked strip', *J. Appl. Mech.* 1966 **33**, 674–675.

(5/37) ISIDA, M. 'On the determination of stress intensity factors for some common strucural problems', *Engg. Fracture Mech.* 1970 **2**, 61–79.

(5/38) ISIDA, M. 'Analysis of stress intensity factors for plates containing random array of cracks', *Bull. JSME* 1970 **13**, 635–642.

(5/39) ISIDA, M. 'Effect of width and length on stress intensity factors of internally cracked plates under various boundary conditions', *Int. J. Fracture Mech.* 1971 **7**, 301–316.

(5/40) ISIDA, M. Proc. Japan-U.S. Seminar: Combined nonlinear and linear fracture mechanics applications to modern engineering structures. To be published.

(5/41) BOWIE, O. L. 'Analysis of an infinite plate containing radial cracks originating at the boundary of an internal circular hole', *J. Math. Phys.* 1956 **25**, 60–71.

(5/42) *Ibid.* (2/11).

(5/43) BOWIE, O. L. and NEAL, D. M. 'Single edge crack in rectangular tensile sheet', *J. Appl. Mech.* 1965 **32**, 708–709.

(5/44) RICH, T. P. and ROBERTS, R. 'Stress intensity factors for plate bending', *J. Appl. Mech.* 1967 **34**, 777–779.

(5/45) BLOOM, J. M. 'The short single edge crack specimen with linearly varying end displacements', *Int. J. Fracture Mech.* 1966 **2**, 597–603.

(5/46) AKAO, H. T. and KOBAYASHI, A. S. 'Stress intensity factor for a short edge-notched specimen subjected to three point loading', *J. Bas. Engg.* 1967 **89**, 7–12.

(5/47) NEAL, D. M. 'Stress intensity factors for cracks emanating from rectangular cutouts', *Int. J. Fracture Mech.* 1970 **6**, 393–400.

(5/48) BOWIE, O. L. and NEAL, D. M. 'A modified måpping-collocation technique for accurate calculation of stress intensity factors', *Int. J. Fracture Mech.* 1970 **6**, 199–206.

(5/49) BOWIE, O. L. and FREESE, C. E. 'Central crack in plane orthotropic rectangular sheet', *Int. J. Fracture Mech.* 1972 **8**, 49–58.

(5/50) BOWIE, O. L. and FREESE, C. E. 'Elastic analysis for a radial crack in a circular ring', *Engg. Fract. Mech.* 1972 **4**, 315–321.

(5/51) BOWIE, O. L., FREESE, C. E. and NEAL, D. M. 'Solution of plane problems of elasticity utilizing partitioning concepts', *J. Appl. Mech.* 1973 **40**, 767–772.

(5/52) NEUBER, H. 'Theory of notch stresses', Springer Berlin 1958.

(5/53) HARRIS, D. O. 'Stress intensity factors for hollow circumferentially notched round bars', *J. Bas. Engg.* 1967 **89**, 49–54.

(5/54) POOK, L. P. and DIXON, J. R. 'Fracture toughness of high-strength materials: theory and practice', *ISI Publ.* 1970 **120**, 45–50.

(5/55) *Ibid.* (1/13).

(5/56) SIH, G. C. and LIEBOWITZ, H. 'Fracture', 2(2), Academic Press, 1968.

(5/57) BARENBLATT, G. I. 'Advances in applied mechanics', **7**, 55–111, Academic Press 1962.

(5/58) EMERY, A. F. and WALKER, G. E. 'Stress intensity factors for edge cracks in rectangular plates with arbitrary loadings', Report No. SCL-DC-67-105, Dept. Mech. Engg., University of Washington, Seattle 1968.

(5/59) CHELL, G. G. 'The stress intensity factor for a crack in a plate subject to an arbitrary stress and with displacement boundary conditions on its ends', Report No. RD/L/N 26/74 Central Electricity Research Laboratories, Leatherhead, Surrey 1974.

(5/60) SNEDDON, I. N. and LOWENGRUB, M. 'Crack problems in the classical theory of elasticity', Wiley 1969.

(5/61) ROOKE, D. P. and SNEDDON, I. N. 'The crack energy and the stress intensity factor for a cruciform crack deformed by internal pressure', *Int. J. Engg. Sci.* 1969 **7**, 1079–1089.

(5/62) TWEED, J. 'The solution of certain triple integral equations involving inverse mellin transforms', *Glasgow Math. J.* 1973 **14**, 65–72.

(5/63) TWEED, J., ROOKE, D. P. and DAS, S. C. 'The stress intensity factor of a radial crack in a finite elastic disc', *Int. J. Engg. Sci.* 1972 **10**, 323–335.

(5/64) ROOKE, D. P. and TWEED, J. 'The stress intensity factors of a radial crack in a finite rotating elastic disc', *Int. J. Engg. Sci.* 1972 **10**, 709–714.

(5/65) TWEED, J. and ROOKE, D. P. 'The torsion of a circular cylinder containing a symmetric array of edge cracks', *Int. J. Engg. Sci.* 1972 **10**, 801–812.

(5/66) SMETANIN, R. J. 'The problem of extension of an elastic space containing a plane annular slit', *PMM* 1968 **32**, 458–462.

(5/67) FICHTER, W. B. 'Stresses at the tip of a longitudinal crack in a plate strip', NASA TR-R-265 1967.

(5/68) ATKINSON, C. 'On dislocation densities and stress singularities associated with cracks and pile ups in homogeneous media', *Int. J. Engg. Sci.* 1972 **10**, 45.

(5/69) ATKINSON, C. 'The interaction between a crack and an inclusion', *Int. J. Engg. Sci.* 1972 **10**, 127–136.

(5/70) HISITANI, H. Proc. Japan—U.S. Seminar: Combined nonlinear and linear fracture mechanics applications to modern engineering structures. To be published.

(5/71) ROMOUALDI, J. P., FRASIER, J. T. and IRWIN, G. R. 'Crack extension force near a riveted stringer', N.R.L. Report 4956, Washington D.C. 1957.

(5/72) BLOOM, J. M. and SANDERS, J. L. 'The effect of a riveted stringer on the stress in a cracked sheet', *J. Appl. Mech.* 1966 **33**, 561–570.

(5/73) KANAZAWA, T., MACHIDA, S. and OHYAGI, M. 'Some basic considerations on crack arresters—Part V', *J. Soc. Naval Arch.* Japan 1967 **122**, 200–214: R.A.E. Library Translation No. 1648, 1972.

(5/74) MADISON, R. B. 'Application of fracture mechanics to bridges', Ph.D. Thesis, Lehigh University 1969.
(5/75) POE, C. C. 'Stress intensity factors for a cracked sheet with riveted and uniformly spaced stringers', NASA Tech. Report R-358 1971.
(5/76) POE, C. C. 'The effect of broken stringers on the stress intensity factor for a uniformly stiffened sheet containing a crack', NASA Tech. Report TMX-71947 1973.
(5/77) VLIEGER, H. 'The residual strength characteristics of stiffened panels containing fatigue cracks', *Engg. Fract. Mech.* 1973 **5**, 447–477.
(5/78) SWIFT, T. 'The effects of fastener flexibility and stiffener geometry on the stress intensity in stiffened cracked sheets', Douglas Paper 6211, McDonnell Douglas Corporation, Long Beach, California. Presented at Conf. on Prospects of Fracture Mechanics, Delft 1974.
(5/79) ARIN, K. 'A plate with a crack, stiffened by a partially debonded stringer', *Engg. Fract. Mech.* 1974 **6**, 133–140.
(5/80) MILLER, M. and CARTWRIGHT, D. J. 'Stress intensity factors for a crack in two intersecting uniformly stressed sheets', *Int. J. Engg. Sci.* 1974 **12**, 353–359.
(5/81) PICARD, E. *'Traite d'analyse'* **2**, 1893.
(5/82) SMITH, F. W., EMERY, A. F. and KOBAYASHI, A. S. 'Stress intensity factors for semicircular cracks', *J. Appl. Mech.* 1967 **34**, 953–959.
(5/83) SMITH, F. W. and ALAVI, M. J. 'Stress intensity factors for a penny shaped crack in a half space', *Engg. Fracture Mech.* 1971 **3**, 241–254.
(5/84) SMITH, F. W. and ALAVI, M. J. 'Stress intensity factors for a part circular surface flaw', *Dept. Mech. Engg.*, Colorado State University.
(5/85) THRESHER, R. W. and SMITH, F. W. 'Stress intensity factors for a surface crack in a finite solid', *J. Appl. Mech.* 1972 **39**, 195–200.
(5/86) SMITH, F. W. 'Stress intensity factors for a semi-elliptical surface flaw', Boeing Company, Structural Development Research Memorandum No. 17. 1966.
(5/87) SHAH, R. C. and KOBAYASHI, A. S. 'Stress intensity factor for an elliptical crack approaching the surface of a plate in bending', *ASTM STP* 513 1972, 3–21.
(5/88) SHAH, R. C. and KOBAYASHI, A. S. 'The surface crack: physical problems and computational methods', *ASME*, New York, 1972.
(5/89) SHAH, R. C. and KOBAYASHI, A. S. 'Stress intensity factors for an elliptical crack approaching the surface of a semi-infinite solid', *Int. J. Fracture Mech.* 1973 **9**, 133–146.
(5/90) ZIENKIEWITZ, O. C. and CHEUNG, Y. K. 'The finite element method in structural and continuum mechanics', McGraw-Hill. 1967.
(5/91) JERRAM, K. and HELLEN, T. K. 'Finite element techniques in fracture mechanics', Proc. Inter. Conf. on Welding Research related to power plant, CEGB, Marchwood Engg. Laboratories. 1972.

(5/92) OGLESBY, J. J. and LOMACKY, O. 'An evaluation of finite element methods for computation of elastic stress intensity factors', *J. Engg. for Industry* 1973 **95**, 177–185.

(5/93) CHAN, S. K., TUBA, I. S. and WILSON, W. K. 'On the finite element method in linear fracture mechanics', Scientific Paper 68-ID7-FMPWR-P1, Westinghouse Research Laboratories, Pittsburgh, Pennsylvania. 1968.

(5/94) KOBAYASHI, A. S., MAIDEN, D. E., SIMON, J. B. and IIDA, S. 'Application of finite element analysis method to two-dimensional problems in fracture mechanics', Publication 69-WA/PVP-12, American Society of Mechanical Engineers. 1969.

(5/95) ANDERSON, G. P., RUGGLES, V. L. and STIBOR, G. S. 'Use of finite element computer programs in fracture mechanics', *Int. J. Fracture Mech.* 1971 **7**, 63–76.

(5/96) WILSON, W. K. and THOMPSON, D. G. 'On the finite element method of calculating stress intensity factors for crack plates in bending', *Engg. Fracture Mech.* 1971 **3**, 97–102.

(5/97) MYAMOTO, H. and MIYOSHI, T. 'Analysis of stress intensity factor for surface-flawed tension plate', presented at the colloquium of IUTAM 'High speed computing of elastic structures', Liege. 1970.

(5/98) *Ibid.* (4/30).

(5/99) *Ibid.* (1/2) 1225–1230, 1958.

(5/100) JERRAM, K. 'The use of a finite element stress analysis computer program for the calculation of stress intensity factors', Report No. RD/B/N1521, Central Electricity Generating Board, Berkeley Nuclear Laboratories. 1970.

(5/101) HELLEN, T. K. 'The calculation of stress intensity factors using refined finite element techniques', Research Report RD/B/N2583. Central Electricity Generating Board, Berkeley Nuclear Laboratories. 1973.

(5/102) SWANSON, S. R. 'Finite element solutions for a cracked two layered elastic cylinder', *Engg. Fracture Mech.* 1971 **3**, 283–289.

(5/103) DIXON, J. R. and POOK, L. P. 'Stress intensity factors calculated generally by the finite element technique', *Nature* 1969 **224**, 166–167.

(5/104) ADAMS, N. J. I. 'The effect of curvature on stress intensity and crack growth in shells', Ph.D. Thesis, University of Southampton. 1969.

(5/105) MOWBRAY, D. F. 'A note on the finite element method in linear fracture mechanics', *Engg. Fracture Mech.* 1970 **2**, 173–176.

(5/106) CARTWRIGHT, D. J. 'Stress intensity factors and residual static strength of certain cracked structural elements', Ph.D. Thesis, University of Southampton. 1971.

(5/107) WILSON, W. K. 'On combined mode fracture mechanics', Research Report 69-1E7-FMECH-R1, Westinghouse Research Laboratories, Pittsburgh, Pennsylvania. 1969.

(5/108) BYSKOV, E. 'The calculation of stress intensity factors using the finite element method with cracked elements', *Int. J. Fracture Mech.* 1970 **6**, 159–167.

(5/109) WALSH, P. F. 'The computation of stress intensity factors by a special finite element technique', *Int. J. Solids Struct.* 1971 **7**, 1333–1341.

(5/110) TRACEY, D. M. 'Finite elements for determination of crack tip elastic stress intensity factors', *Engg. Fracture Mech.* 1971 **3**, 255–265.

(5/111) PIAN, T. H. H., TONG, P. and LUK, C. H. 'Elastic crack analysis by a finite element hybrid method', Proceedings 3rd Airforce Conf. on Matrix Methods in Structural Mechanics, Wright Patterson Base, Ohio, U.S.A. 1971.

(5/112) RAO, A. K., RAJU, I. S. and KRISTNA MURTY, A. V. 'A powerful hybrid method in finite element analysis', *Int. J. Num. Methods Engg.* 1971 **3**, 389–403.

(5/113) YAMAMOTO, Y. 'Finite element approaches with the aid of analytical solutions', Presented at the Japan—U.S. Seminar on Matrix Methods of Structural Analysis and Design, Tokyo. August, 1969.

(5/114) BLACKBURN, W. S. and HELLEN, T. K. 'Calculation of stress intensity factors for elliptical and semi-elliptical cracks in blocks and cylinders', Research Report RD/B/N3103, Central Electricity Generating Board, Berkeley Nuclear Laboratories. 1974.

(5/115) BLACKBURN, W. S. 'Calculation of stress intensity factors at crack tips using special finite elements', Conf. on Maths of Finite Elements at Crack tips using special elements, Brunel University. 1972.

(5/116) IRWIN, G. R. and KIES, J. A. 'Critical energy rate analysis of fracture strength', *Weld. J.* (Res. suppl.) 1954 **33**, 193s–198s.

(5/117) Ibid. (1/12).

(5/118) HUCULAK, P. 'Stress intensity factors for a crack progressing along a series of equally spaced holes', Aeronautical Report, M.S. 117, Canada. 1968.

(5/119) CARTWRIGHT, D. J. and RATCLIFFE, G. A. 'Strain energy release rate for radial cracks emanating from a pin loaded hole', *Int. J. Fracture Mech.* 1972 **8**, 175–181.

(5/120) UNDERWOOD, J. H., LASSELE, R. R., SCANLON, R. D. and HUSSAIN, M. A. 'A compliance $K$ calibration for a pressurised thick-wall cylinder with a radial crack', *Engg. Fracture Mech.* 1972 **4**, 231–244.

(5/121) GALLAGHER, J. P. 'Experimentally determined stress intensity factors for several contoured double cantilever beam specimens', *Engg. Fracture Mech.* 1971 **3**, 27–43.

(5/122) KOBAYASHI, A. S. (Ed.) 'Experimental techniques in fracture mechanics', SESA Monograph. 1973.

(5/123) SCHROEDL, M. A., McGOWAN, J. J. and SMITH, C. W. 'An assessment of factors influencing data obtained by the photoelastic stress freezing technique for stress fields near crack tips', *Engg. Fract. Mech.* 1972 **4**, 801–809.

(5/124) HARDY, A. K. 'Determination of stress intensity factors: an evaluation of two methods', M.Sc. Thesis, University of Bath 1972.

(5/125) MARLOFF, R. H., LEVEN, M. M., RINGLER, T. N. and JOHNSON, R. L. 'Photoelastic determination of stress intensity factors', *Expl. Mech.* 1971 **11**, 529–539.

(5/126) CREAGER, M. 'The elastic stress field near the tip of a blunt crack', M.Sc. Thesis, Lehigh University, U.S.A. 1966.

(5/127) EMERY, A. F., BARRETT, C. F. and KOBAYASHI, A. S. 'Temperature distributions and thermal stresses in a partially filled annulus', *Expl. Mech.* 1966 **6**, 602–608.

(5/128) SMITH, D. G. and SMITH, C. W. 'A photoelastic evaluation of the influence of closure and other effects upon the local bending stresses in cracked plates', *Int. J. Fracture Mech.* 1970 **6**, 305–318.

(5/129) SMITH, D. G. and SMITH, C. W. 'Photoelastic determination of mixed mode stress intensity factors', *Engg. Fract. Mech.* 1972 **4**, 357–366.

(5/130) WILSON, W. K. 'Comparison of the stress distribution on the plane of symmetry at the WOL test specimen obtained by various methods and an interpretation of the results of a photoelastic study of the specimen', Research Report 66-1D7-MEMTL-R2, Westinghouse Research Laboratories, Pittsburgh, Pennsylvania. 1966.

(5/131) LEVEN, M. M. 'Stress distribution in the M4 biaxial fracture specimen', Research Report 65-1D7-STRSS-R1, Westinghouse Research Laboratories, Pittsburgh, Pennsylvania. 1965.

(5/132) PARIS, P. C. 'Fatigue—An interdisciplinary approach', 107–127, Eds. J. J. Burke, N. L. Reed and V. Wiess, Syracuse University Press. 1964.

(5/133) JAMES, L. A. and ANDERSON, W. E. 'A simple experimental procedure for stress intensity factor calibration', *Engg. Fracture Mech.* 1969 **1**, 565–568.

(5/134) DUDDERAR, T. D. and GORMAN, H. J. 'The determination of mode I stress intensity factors by holographic interferometry', *Expl. Mech.* 1973 **13**, 145–149.

(5/135) SOMMER, E. 'An optical method for determining the crack-tip stress intensity factor', *Engg. Fracture Mech.* 1970 **1**, 705–718.

(6/1) SRAWLEY, J. E. and BROWN, W. F. 'Fracture toughness testing methods', ASTM STP 381.

(6/2) ASTM Special Committee on Fracture testing of High Strength Metallic Materials 'Progress in measuring fracture toughness using fracture mechanics', Materials Research and Standards, **4** (3) March 1964, 107–119.

(6/3) BOYLE, R. W., SULLIVAN, A. M. and KRAFFT, J. M. 'Determination of plane strain toughness with sharply notched sheets', Welding Journal (Research Supplement) **41**, 9, September 1962, 428s–432s.

(6/4) LINDLEY, C., WILKINSON, T. and PRIEST, A. H. 'A clip gauge extensometer for use at elevated and cryogenic temperatures', British Steel Corporation, Physical Metallurgy Centre, Technical Note PMC/APE/23/74.

(6/5) Ibid. (2/22).

(6/6) MAY, M. J. 'British experience with plane strain fracture toughness ($K_{Ic}$) testing', ASTM STP 463, 1970, 41–62.

(6/7) WILKINSON, T. and WALKER, E. F. 'The influence of crack length and thickness on plane strain fracture toughness tests', British Steel Corporation, Corporate Laboratories Report, MG/30/72.

(6/8) Ibid. (5/24).

(6/9) WALKER, E. F. and MAY, M. J. 'Compliance functions for various types of test specimen geometry', BISRA Report No. MG/E/307/67.

(6/10) PRIEST, A. H. and MAY, M. J. 'Fracture toughness testing in impact', BISRA Report MG/C/46/69.

(6/11) VENZI, S., PRIEST, A. H. and MAY, M. J. 'Influence of inertial load in instrumented impact tests', ASTM STP 466, 165–180.

(6/12) VALERIANI, G. and VENZI, S. 'Determination of $K_{Ic}$ and COD in impact tests', Centro Sperimentale Metallurgico Report for ECSC, 1974.

(6/13) IRELAND, D. R. 'Procedures and problems associated with reliable control of the instrumented impact test' *ASTM STP* 563, 3–29.

(6/14) WELLS, A. A. 'The status of COD in fracture mechanics', Canadian Congress of Applied Mechanics, 1971.

(6/15) INGHAM, T. et al. 'The effect of geometry on the interpretation of COD test data', Proceedings of Conference on Practical Applications of Fracture Mechanics to Pressure Vessel Technology, Institution of Mechanical Engineers, London, May 1971.

(6/16) VENZI, S. 'Determination of a generalized relationship for COD calibration', Centro Sperimentale Metallurgico Report 1973.

(6/17) BARR, R. R., ELLIOTT, D., TERRY, P. and WALKER, E. F. 'The measurement of COD and its application to defect significance', British Steel Corporation Report GSD/PMC/1/75.

(6/18) PRIEST, A. H. and MAY, M. J. 'Strain rate transitions in structural steel', BISRA Report MG/C/95/69.

(6/19) McINTYRE, P. and PRIEST, A. H. 'Measurement of sub-critical flaw growth in stress corrosion, cyclic loading and high temperature creep by the DC electrical resistance technique', BSC Corporate Laboratories Report MG/54/71.

(6/20) McINTYRE, P. and ELLIOTT, D. 'The role of stretch zone formation in environmentally activated crack growth in steels', Proc. Third International Conference on the Strength of Metals and Alloys, Cambridge 1973.
(6/21) KNOTT, J. 'The fracture toughness of metals', Chapter 3, this work.
(6/22) 'Proposed recommended standard for $R$-curve determination', Attachment 2, ASTM E24-01 Minutes, March 7, 1973.
(6/23) JUDY, R. W. and GOODE, R. J. 'Fracture extension resistance ($R$-curve) characteristics for three high strength steels', ASTM STP 527, 48–61.
(6/24) HOLMES, B. and PRIEST, A. H. 'Shear fracture propagation resistance of linepipe steels', *BSC-TNO Report for ECSC/EPRG Project 6210*. 46/6/601, 1976.
(6/25) LANDES, J. D. and BEGLEY, J. A. 'The effect of specimen geometry on $J_{Ic}$', ASTM STP 514, 24–39.
(6/26) LOGSDON, W. A. 'Elastic plastic ($J_{Ic}$) fracture toughness values: their experimental determination and comparison with conventional linear elastic ($K_{Ic}$) fracture toughness values for five materials', Westinghouse Electric Corporation, Scientific Paper 74-1E7-FMPWR-P1, 1974.
(6/27) RODDIS, R. and PRIEST, A. H. 'An investigation of Rice's $J$ integral as a fracture criterion', British Steel Corporation Report PMC/6791/-/74/A.
(6/28) WITT, F. J. 'Equivalent energy procedures for predicting gross plastic fracture', Oak Ridge National Laboratory Paper ORNL-TM-3172.
(6/29) BUCHALET, C. and MAGER, T. R. 'Experimental verification of lower bound $K_{Ic}$ values utilising the equivalent energy concept', ASTM STP 536, 281–296.
(6/30) ORNER, G. M. and HARTBOWER, C. E. 'Sheet fracture toughness evaluated by Charpy impact and slow bend', Welding Journal (Research Supplement) **40**, September 1961, 405s–416s.
(6/31) Draft Standard for Study 50B 'Pre-crack Charpy test method', AECMA C5 Study 50.
(7/1) 'Report of royal commission into the failure of Kings Bridge', Presented to Parliament (of) Victoria, Melbourne 1963.
(7/2) SMITH, N. and HAMILTON, I. G. West of Scotland Journal **76**, 1968/69. Paper 591, 111–153.
(7/3) SAMONS, C. H. API meeting Houston, Texas, May 1954.
(7/4) PARKER, E. R. 'Brittle behaviour of engineering structures', John Wiley and Sons, New York, 1957.
(7/5) HODGSON, J. and BOYD, G. M. Trans. Royal Institution of Naval Architects, 1958 **100**, 141–180.
(7/6) ROBERTSON, T. S. *JISI* 1953 **175**, 361.
(7/7) PELLINI, W. S. 'Criteria for fracture Control', NRL Report 7406, May, 1972.
(7/8) WELLS, A. A. *Welding Journal* 1961 **26**, 4, 182s–192s.

(7/9)   Anon. *Brit. Weld. J.* 1955 **2**, 6, 254–263.
(7/10)  GREENE, T. W. *Welding Journal* 1949 **14**, 193s.
(7/11)  WOODLEY, C. C., BURDEKIN, F. M. and WELLS, A. A. *Brit. Weld. J.* 1964 **11**, 3, 123–136.
(7/12)  ROSE, R. T. BWRA Bulletin, May 1962.
(7/13)  British Standards Institute BS 1515: Part 1: 1965. Appendix C.
(7/14)  British Standards Institute BS 4741, 1971.
(7/15)  British Standards Institute BS 2654, 1974.
(7/16)  *Ibid.* (1/1) 163–198.
(7/17)  IRWIN, G. R. U.S. Naval Research Lab. Report NRL 4763–1956.
(7/18)  *Ibid.* (4/16).
(7/19)  COTTRELL, A. H. 'Steels for reactor pressure circuits', 281–296, 1960, ISI.
(7/20)  American Society for Testing and Materials ASTM S7P381. 'Plane strain crack toughness testing of high strength metallic materials', 1967.
(7/21)  KIES, J. A., SMITH, H. L. and STONESIFER, F. R. NRL Report 6918, Sept. 1969.
(7/22)  *Ibid.* (3/7).
(7/23)  *Ibid.* (3/31).
(7/24)  DAWES, M. G. *Welding Journal* 1974 **53**, 9, 369s–379s.
(7/25)  BURDEKIN, F. M. IIW Document X-749-74, 'Report of working group on significance of defects'—Published.
(7/26)  BAKER, R. G., BARR, R. R., GULVIN, T. F. and TERRY, P. 2nd Int. Conf. on Pressure Vessel Technology, San Antonio, 1973.
(7/27)  DAWES, M. G. Welding Research International, 1974 **4**, 4, 41–73.
(7/28)  BARR, R. R., ELLIOTT, D., TERRY, P. and WALKER, E. F. BSC Report GSD/PMC/1, 1975. 'The measurement of COD and its application to defect significance'.
(7/29)  DAWES, M. G. Met Construction and Brit. Weld. J. 1971 **3**, 2, 61.
(7/30)  DOLBY, M. A. and ARCHER, G. L. *Proc. I. Mech. E. Conf.*, London, May 1971.
(7/31)  DOLBY, M. A. Met Construction and Brit Weld. J. 1974 **6**, 228–233.
(7/32)  BARR, R. R. and BURDEKIN, F. M. 'Design and brittle fracture' Rosenham Centenary Conference, NPL. Sept. 1975.
(7/33)  HAYES, D. J. and MADDOX, S. J. Welding Institute Research Bulletin, **13** (1), January 1972.
(7/34)  BURDEKIN, F. M. Welding Institute Autumn Meeting Conference, November 1975.
(8/1)   FROST, N. E., MARSH, K. J. and POOK, L. P. *Metal Fatigue* 1974. (Clarendon Press, Oxford).
(8/2)   MOORE, H. F. and KOMMERS, J. B. *The fatigue of metals* 1927. (McGraw-Hill, New York).

(8/3) FROST, N. E. 'The current state of the art of fatigue: its development and interaction with design'. Society of Environmental Engineers Fatigue Group Conference on the World of Fatigue. City University, London. (To be published in J. Soc. Env. Engrs.)

(8/4) FORSYTH, P. J. E. A two-stage process of fatigue crack growth. *Proc. Crack Propagation Symp.*, 1961, 76–94, Cranfield College of Aeronautics, 1962.

(8/5) WEBER, J. H. and HERTZBERG, R. W. 'Effect of thermomechanical processing on fatigue crack propagation', *Met. Trans.* 1973 **4**(2), 595–601.

(8/6) ZWEBEN, C. 'On the strength of notched composites', *J. Mech. Phys. Solids* 1971 **19**(3), 103–116.

(8/7) POOK, L. P. and HOLMES, R. 'Bi-axial fatigue crack growth tests'. To be presented at the Society of Environmental Engineers Fatigue Group, Fatigue Testing and Design Conference and Exhibition. The City University, London, 5–9 April, 1976.

(8/8) ARAD, S., RADON, J. C. and CULVER, L. E. 'Design against fatigue failure in thermoplastics', *Engng fract. Mech.* 1972 **4**(3), 511–522.

(8/9) FROST, N. E., POOK, L. P. and DENTON, K. 'A fracture mechanics analysis of fatigue crack growth data for various materials', *Engng fract. Mech.* 1971 **3**, 109–126.

(8/10) LINDLEY, T. C. and RICHARDS, C. E. 'The relevance of crack closure to fatigue crack propagation', *Mater. Sci. & Engng* 1974 **14**(3), 281–293.

(8/11) HOEPPNER, D. W. and KRUPP, W. E. 'Prediction of component life by application of fatigue crack growth knowledge', *Engng fract. Mech.* 1974 **6**(1), 47–70.

(8/12) POOK, L. P. 'Fracture mechanics; how it can help engineers', *Trans. N-E Coast Inst. Engrs and Shpbldrs* 1974 **90**(3), 77–92.

(8/13) POOK, L. P. 'Basic statistics of fatigue crack growth'. Paper presented at a symposium on Statistical Aspects of Fatigue Testing held by the Society of Environmental Engineers Fatigue Group at Warwick University on 12 Feb. 1975. (To be published in *J. Soc. Env. Engrs.*)

(8/14) CLARK, W. G. and HUDAK, S. J. 'Variability in fatigue crack growth rate testing', ASTM E-24.04.01 Task Group Report. Westinghouse Research Laboratories. Scientific Paper 74-IE7-MSLRA-P2, 1974.

(8/15) POOK, L. P. and GREENAN, A. F. *Fatigue crack growth threshold for mild steel, a low alloy steel and a grey cast iron* 1974. *NEL Report* 571. (National Engineering Laboratory, East Kilbride, Glasgow.)

(8/16) SCHIJVE, J. 'Fatigue damage accumulation and incompatible crack front orientation', *Engng fract. Mech.* 1974 **6**(2), 245–252.

(8/17) CARDEN, A. E., McEVILY, A. J. and WELLS, C. H. (Eds). 'Fatigue at elevated temperature', *ASTM STP* 520, 1973. (American Society for Testing and Material, Philadelphia PA.)

(8/18)  GRAY, I. and HEATON, M. D. 'Application of fracture mechanics to the fatigue of power plant components'. *Designing against Fatigue—Implications of Recent Findings on Complex Situations* 1974. (IMechE., London.)

(8/19)  POOK, L. P. 'Fracture mechanics analysis of the fatigue behaviour of welded joints', *Weld. Res. Int.* 1974 **4**(3), 1–24.

(8/20)  GURNEY, T. R. and MADDOX, S. J. 'Determination of fatigue design stresses for welded structures from an analysis of data', *Metal Constn* 1972 **4**(11), 418–422.

(8/21)  WATSON, P. and DABELL, B. J. 'Cycle counting and fatigue damage', Paper presented at a symposium on Statistical Aspects of Fatigue Testing held by the Society of Environmental Engineers Fatigue Group at Warwick University on 12 Feb. 1975. To be published in *J. Soc. Env. Engrs*.

(8/22)  WHEELER, O. E. 'Spectrum loading and crack growth', *J. bas. Engng* 1972 **94D**(1), 181–186.

(8/23)  BRUSSAT, T. R. 'An approach to predicting the growth to failure of fatigue cracks subjected to arbitrary uniaxial cyclic loading'. *Damage tolerance in aircraft structures*. ASTM STP 486, 122–147, 1971. American Society for Testing and Materials, Philadelphia PA.

(8/24)  MARSH, K. J. 'Full-scale testing—an aid to the designer', *J. Soc. Env. Engrs* 1974 **13–4**, 63, 15–16, 21–22.

(8/25)  SCHÜTZ, W. 'Fatigue life prediction of aircraft structures—past, present and future', *Engng fract. Mech.* 1974 **6**(3), 671–699.

(9/1)  PARKINS, R. N. *British Corrosion Journal* 1972 **7**, 15.

(9/2)  WEST, J. M. *Metal Science Journal* 1973 **7**, 169.

(9/3)  PARKINS, R. N. Proc. Conf. on 'Stress corrosion cracking and hydrogen embrittlement of iron base alloys'. To be published. National Association of Corrosion Engineers, Houston, U.S.A.

(9/4)  UHLIG, H. H. 'Fundamental aspects of stress corrosion cracking', Edited by R. W. Staehle, A. J. Forty and D. van Rooyen, 1969, NACE, Houston, U.S.A. 86.

(9/5)  WIEDERHORN, S. M. 'Corrosion fatigue: chemistry, mechanics and microstructure'. Edited by O. F. Devereux, A. J. McEvily and R. W. Staehle, 1971, NACE, Houston, U.S.A. 731.

(9/6)  KAMBOUR, R. P. loc. cit. 681.

(9/7)  PARKINS, R. N. 'The interactions of stress and corrosion in the stress corrosion cracking of iron base alloys'. Paper 6, Proc. Conf. Mechanics and Mechanisms of Fracture, Churchill College, Cambridge, 1973.

(9/8)  ORIANI, R. A. Proc. Conf. 'Stress corrosion cracking and hydrogen embrittlement of iron base alloys'. To be published. NACE, Houston, U.S.A.

(9/9)  ST. JOHN, C. and GERBERICH, W. W. Met Trans. 1973 **4**, 589.

(9/10)  SPEIDEL, M. O. Conference on 'Hydrogen in metals', to be published, 1975, NACE.

(9/11) SPEIDEL, M. O. 'The theory of stress corrosion cracking in alloys', Edited by J. C. Scully, NATO, Brussels, 1971, 289.
(9/12) FEENEY, J. A. and BLACKBURN, M. J. loc. cit. 355.
(9/13) BROWN, B. F. and BEACHAM, C. D. *Corrosion Science* 1965 **5**, 745.
(9/14) WEARMOUTH, W. R., DEAN, G. P. and PARKINS, R. N. *Corrosion* 1973 **29**, 251.
(9/15) SYRETT, B. C. *Corrosion* 1971 **27**, 270.
(9/16) PARKINS, R. N. 5th Symposium on Line Pipe Research, American Gas Association, Catalogue No. L.30174, 1974.
(9/17) McINTYRE, P. Proc. Conf. 'Stress corrosion cracking and hydrogen embrittlement of iron base alloys'. To be published. NACE, Houston, U.S.A.
(9/18) POURBAIX, M. 'The theory of stress corrosion cracking in alloys'. Edited by J. C. Scully, NATO, Brussels, 1971, 17.
(9/19) SMITH, J. A., PETERSON, M. H. and BROWN, B. F. *Corrosion* 1970 **26**, 539.
(9/20) CONGLETON, J. Proc. Conf. 'Stress corrosion cracking and hydrogen embrittlement of iron base alloys'. To be published. NACE, Houston, U.S.A.
(10/1) BUNTIN, W. D. Concept and conduct of proof test of F-111 production aircraft. *Aeronaut. J.* 1972 **76**(742), 587–598.

# ADDENDUM

*Additional References from Chapter* 4.
(5/A) CHELL, G. G. and DAVIDSON, A. 'A post-yield fracture mechanics analysis of single edge notched tension specimens', *Mat. Sci. and Engng* 1976, **24**, 45–52
(5/B) CHELL, G. G. 'The stress intensity factors and crack profiles, for centre and edge cracks in plates subject to arbitrary stresses', *Int. J. Fract.* 1976, **12**, 33–46.
(5/C) CHELL, G. G. 'The application of post-yield fracture mechanics to penny shaped and semi-circular cracks', *Eng. Fract. Mech.* 1977, **9**, 55–64.
(5/D) PERIS, P. C. *et al.* 'A treatment of the subject of tearing instability'. Symposium on elastic plastic fracture. Atlanta, November 1977. To be published, ASTM.
(5/E) SUMPTER, J. D. G. and TURNER, C. E. 'Use of the $J$ contour integral in elastic–plastic fracture studies by finite-element methods', *J. Mech. Eng. Sci.* 1976, **18**, 97–112.
(5/F) TURNER, C. E. 'An analysis of the fracture implications of some elastic–plastic finite element studies'. *Numerical Methods in Fracture Mechanics*, Ed. Luxmoore and Owen, University College, Swansea, 1978, 569–580.
(5/G) McMEEKING, R. M. and PARKS, D. M. 'A criterion for $J$-dominance of crack tip fields in large scale yielding, *ibid* (5/D).

(5/H)  GARWOOD, S. J., ROBINSON, J. N. and TURNER, C. E. 'The measurement of crack growth resistance curves ($R$-curves) using the $J$ integral', *Int. J. Fract.* 1975, **11**, 528–531.

(5/I)  GARWOOD, S. J. The measurement of crack growth resistance using yielding fracture mechanics. Ph.D. Thesis, University of London, December 1976.

(5/J)  GARWOOD, S. J. and TURNER, C. E. 'The use of the $J$ integral to measure the resistance of mild steel to slow stable crack growth', *Fracture* 1977 ICF 4 University of Waterloo, Ed. Taplin, 1977, **2**, 279–284.

(5/K)  TURNER, C. E. 'Description of stable and unstable crack growth in the elastic–plastic regime in terms of $J_r$ resistance curves'. *Eleventh National Symposium on Fracture.* Virginia 1978. To be published ASTM.

(5/L)  MILNE, I. and CHELL, G. G. 'Effect of size on the $J$ fracture criterion', *ibid* (**5/D**).

# Index

Alternating methods, 66

BARR, R. R. *Paper* Application of fracture mechanics to the brittle fracture of structural steels, 93
Boundary collocation, method of, 58
BRITTLE FRACTURE OF STRUCTURAL STEELS, APPLICATION OF FRACTURE MECHANICS TO, *Paper*, 93

CARTWRIGHT, D. J. *Paper* Evaluation of stress intensity factors, 54
Compliance, 69
Conformal mapping, 61
Crack growth data, fatigue, analysis and application of, 114
Crack growth, environmental effects in, 136
Crack opening displacement (COD), 21, 37, 41, 51, 83
Crack tip elements, 68
Crack tip stress and displacement, 67

Dislocation models, 63

Elastic/plastic stress states, 21
Electrochemical effects in geometrical discontinuities, 151
Energy balance approach to fracture, 138
ENERGY BALANCE APPROACH TO FRACTURE, ORIGINS OF, *Paper*, 1
Environmental effects in crack growth, 136
Equivalent energy procedure, 89

FATIGUE CRACK GROWTH DATA, ANALYSIS AND APPLICATION OF, *Paper*, 114
Fatigue crack growth rate, 71

Finite-elements, determining stress intensity factors using, 66
Force-displacement matching, 65
Fracture, energy balance approach to, 138
Fracture, energy balance approach to, origins of, 1
Fracture mechanics, analyis and application of fatigue crack growth data, 114
Fracture mechanics, application to brittle fracture of structural steels, 93
Fracture mechanics, application to design and material selection, 109
Fracture mechanics, environmental effects in crack growth, 136
Fracture mechanics, fracture toughness of metals, 17
Fracture mechanics, general notation, vii
Fracture mechanics, origins of the energy balance approach to fracture, 1
Fracture mechanics, origins of the stress intensity factor approach to fracture, 9
Fracture mechanics, summary of its aims and methods, 153
Fracture mechanics, yielding fracture mechanics, 32
Fracture mechanics tests, 101
Fracture mechanics, usage of, 153
Fracture, micro-mechanism of, 26
Fracture, stress intensity factor approach to, origins of, 9
FRACTURE TOUGHNESS, EXPERIMENTAL METHODS FOR MEASUREMENT OF, *Paper*, 74
FRACTURE TOUGHNESS OF METALS, *Paper*, 17

Green's functions, 62

HAYES, D. J. *Papers* Origins of the energy balance approach to fracture, 1. Origins of the stress intensity factor approach to fracture, 9
Holography, 72

Impact fracture propagation energy, 91
Integral transforms, 63
Interferometry, 72

J integral, 41, 46, 51, 89

KNOTT, J. F. *Paper* The fracture toughness of metals, 201

PARKINS, R. N. *Paper* Environmental effects in crack growth, 136
Photoelasticity, determining stress intensity factors by, 69
POOK, L. P. *Paper* Analysis and application of fatigue crack growth data, 114
PRIEST, A. H. *Paper* Experimental methods for fracture toughness measurement, 74

ROOKE, D. P. *Paper* Evaluation of stress intensity factors, 54

Slip line fields, 34
Strain rate effects in crack growth, 144
Stress concentration factor, 54
Stress concentrations, 61
Stress corrosion cracking and stress intensity factor, 143
Stress corrosion crack propagation models, 139
Stress corrosion cracks morphology, 151
Stress intensity factor and stress corrosion crack velocity, 143
STRESS INTENSITY FACTOR APPROACH TO FRACTURE, ORIGINS OF, *Paper*, 9
STRESS INTENSITY FACTORS, EVALUATION OF, *Paper*, 54
Structural steels, application of fracture mechanics to the brittle fracture of, 93

TERRY, P. *Paper* Application of fracture mechanics to the brittle fracture of structural steels, 93

TURNER, C. E. *Paper* Yielding fracture mechanics, 32

William's stress function, 58

YIELDING FRACTURE MECHANICS, *Paper*, 32